The
# Numberverse

How numbers are bursting out of
everything and just want to have fun

## Andrew Day

Edited by Peter Worley
The Philosophy Foundation

Crown House Publishing Limited
www.crownhouse.co.uk

First published by

Crown House Publishing Ltd
Crown Buildings, Bancyfelin, Carmarthen, Wales, SA33 5ND, UK
www.crownhouse.co.uk

British Library Cataloguing-in-Publication Data
A catalogue entry for this book is available
from the British Library.

Print ISBN 978-184590889-8
Mobi ISBN 978-184590897-9
ePub ISBN 978-184590898-0
ePDF ISBN 978-184590899-7

Printed and bound in the UK by
TJ International, Padstow, Cornwall

For Mick Day, who taught me almost all of this without me noticing.

# Foreword by Robin Ince

We are surrounded by numbers; numbers to persuade us, numbers to scare us, numbers to categorise us.

Political parties try to lure us into their way of thinking with big numbers that are meant to sum up wealth or sickness. Soap and yoghurt advertisements try to attract us with numbers that aim to persuade us our guts would be happier or our faces more beautiful if we purchase them.

We are numbers. We exist, knowingly or unknowingly – from our bank account to our national insurance number – as numbers in files, documents and programmes. I have no idea how many different numbers I am. Do you know how many you are? I am 45, I live at 81, my shoes are size 9 and I am often on the 1046.

Despite their omnipresence, numbers can still jar our mind and befuddle our judgement. They can make reality seem unreal.

Numbers are a language we use to explain the world – a language that confounds many of us, too. With the enormous enlightening and obfuscating fog of the mass media, the flood of information now available to us, it is a language we must be willing to interrogate if we are to understand ourselves and the world we live in.

Andrew Day offers the first steps to opening the minds of children to the excitement of numbers and why, through centuries of civilisation and human imagination, they have helped us to frame and shape our world.

These are the numbers that help us to understand why the universe is as it seems to be.

These are the numbers that are pored over by physicists watching bundles of particles colliding at speeds near that of light.

These are the numbers that tell us what page to turn to.

When I think of numbers I think of Douglas Adams and 42, the Answer to Life, the Universe and Everything.

I think of 150, Dunbar's number, the number of people you can maintain a stable social relationship with before your social brain starts to struggle.

I think of 12, the number of angry men in one of my favourite childhood films.

And I think of 23, the numbers of pairs of chromosomes that make us what we are, a quizzical creature that thrives on its own curiosity.

The delight of being a parent is that it gives you an alibi to interrogate the world in the guise of helping your child to learn. This act of educational altruism offers not merely the joyful reward of seeing a child comprehending the world, but also revivifying the adult's intrigue in it all too.

This book reminds us that it is not enough to teach the basics of mathematics. We also need to ask why maths exists and what a world without it might be like. I believe that by understanding why something exists or was created, we are further drawn into a subject. It is not just a case of knowing your times tables but understanding the needs that led to their creation.

The more we understand why, the more we are drawn into the adventure of learning.

# Contents

Contents

# Introduction

## How This Book Was Born

I didn't mean to get interested in maths. I wasn't even supposed to be doing it. What I normally do is visit schools and run philosophy sessions for an hour at a time. Doing philosophy with children doesn't mean telling kids the names of old dead men and what their theories were. We get children to philosophise themselves: to give reasons and compare ideas. We call these lessons enquiries. In our class enquiries we touch on, among other things, science, morals, religion, language … and sometimes numbers.

Then one day a school asked me to show their staff how to use the philosophical approach with all subjects – in other words, to turn lessons into enquiries, putting the children themselves at the heart of the lesson, with their curiosity driving it. This was daunting because I was training people more qualified and experienced than me. But at least I had an idea. So, with the help of some colleagues and the teachers themselves, I began to make some progress. That is how I ended up trying to create maths lessons.

Teachers are sometimes surprised to be told that philosophy and mathematics are closely related. They expect there to be links between philosophy and religion, psychology and literature, and there are, but philosophy has been practised by mathematicians, and vice versa, from the very beginning. In fact, Pythagoras, one of the earliest mathematicians, and perhaps the easiest one to name, was also a philosopher in the sense we understand it now.

As a philosopher, I had a grasp of some of the fundamental questions of what mathematics is. Admittedly, having not studied maths to a very high level it was impossible for me to follow the expert answers to those questions. But, as I explained to some of the teachers I met, not being good at something is sometimes an ideal position from which to teach it, because you can empathise with the students, whereas if you find something easy, your hardest task can be to see why someone else would fail to understand.

When I sat down to start, my first thought was that you would need to make mathematics practical – to show how the things being taught are useful in everyday life. But, as I read the curriculum documents, I found constant references to children acquiring a sense of 'number' – in the abstract. They were supposed to gain an understanding of how the whole number system fitted together and the patterns that run through it. Good luck with that, I thought. I was sure children wouldn't be interested unless you found a practical application.

How wrong I was. I don't think I've ever been more wrong about anything.

As soon as I set youngsters puzzles using pure numbers they leapt into them. And what's more, they had questions. And theories. Sometimes they were well wide of the mark, but because I didn't mind, neither did they. Time after time, I saw children trying to make more sense of what they already knew, and to connect it up to new ideas. To them, the most elementary bits of mathematics were open to question and nothing was taken as read. I persevered with the practical aspect as well, because I thought it was important, but it often took more effort to engage the children that way.

Gradually, I started to understand how the way we run philosophy lessons can be a valuable tool in itself, and there have been moments where a whole class has suddenly been gripped by a question that they themselves have come up with: is zero divisible by anything? Is a square a rectangle? Is 7 a digit or a number? How many lines of symmetry does a square have? What about a circle? And I'm talking about children as young as 6, and in lower ability groups.

Why does it matter if moments like this arise? Well, first of all, we are always in a better position if children are *happy* to be doing maths, even occasionally. Whereas if they know there is no chance of them ever enjoying a maths lesson, they will switch off before it even starts.

I know what that feels like. I found maths dreary when I was a kid. We did maths in blue books (English was in warm, friendly orange or red) and whenever the stack of blue books came out, my heart sank. In maths, you got told about a thing, like long division, and then you did it over and over again in the hope that one day you would stop making mistakes. If you did stop making mistakes, you had to wait until other people stopped too before you could do something else.

That is probably slightly unfair on my teachers, but that is pretty much how I remember it. And you could rightly argue that things have got better since the early 1980s. However, there are still a lot of teachers out there who had the same negative experience as me, and they can't help passing some of that anti-maths feeling on to their own classes. Crucially, they learned to judge themselves as bad at (or bored by) maths very early on, and, as adults, the moment they get something wrong, or can't solve something, they are not intrigued but repelled.

Despite my complaints, I was very lucky because I have a few very different memories to draw on. Here is one …

Aged about 11, my class had two visitors from the local university one morning a week: a short bald man and a tall woman. They showed us how to do algebra and draw graphs of the equations we made. We made up our own equations and challenged each other to solve them. It was pure fun, pure exploration. It was nothing like a maths lesson. When I did algebra in secondary school three years later I lapped it up, probably because I was biased in its favour.

Another oasis in the desert of boredom taught me something else: the value of a sense of purpose. It is the reason why the question, 'Miss, why are we doing this?', is actually a good one most of the time.

It was one morning with Mr Williams when we were 14. We had done trigonometry the year before, and I had hated it, completely alienated by the fact that we had to look up logarithms in the back of a little book. As far as I was concerned, if you were going to look stuff up in a book, why not just look up the answer! Those logarithms meant nothing to me, and I was the kind of kid who needed to know why something was done. Then one day Mr Williams showed us how you could, if you wanted, calculate any of those logarithms yourself. I don't remember how it was done (on the board, I remember, there was a circle cut into quarters and different quarters related to sine, cosine or tangent, and the fourth wasn't used) but remembering the details wasn't the point of the exercise. The point was that we now saw why we don't calculate those values each time – because it takes ages.

When I visit classrooms now, these two experiences influence what I want to bring in with me: the sense of *pleasure* we had when exploring algebra, and the sense of *purpose* I got from having trigonometry procedures explained.

With this in mind, I have looked at a number of mathematical topics, and wondered how to give people who study them both pleasure and purpose. And my best answers fall into

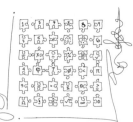

two categories: puzzles, which we do for pleasure, and problems, which we solve for a purpose.

A puzzle is something that tickles our brain and exercises it, but is not useful. Crosswords, for example, are of no value in the struggle to survive, and jigsaws are just time-consuming ways of making a picture that we could buy whole. But they are proof that we can enjoy exercising our brains just for the fun of it.

A problem, on the other hand, I would define as something that we feel driven to fix, even if that problem is an imaginary scenario. Examples of this would be how much interest a debt will rack up over a given period, or how many pounds you have to lose per week if you want to be half a stone thinner for your holiday.

The use that problems and puzzles have is that they are compelling. Once we see the problem, our minds can't help but try to solve it. It is just how we are. And that is the hook, the spark and trigger that a lesson needs. If children are guided, and not merely instructed, they will become adept at recognising problems and engaging with them.

It is really important to remember that children aren't born hating maths. They learn that hate. What they are born with is the impulse to enjoy exploring ideas and make sense of what they are doing. We just need to tap into it, and not trample on it because our eyes are fixed on assessments, streaming and objectives. I'm not against those three things by the way – I just want them not to spoil teaching.

Far too often, the need to simply get the answers right will take over. Then, children who can pick up new methods quickly will excel, and anyone sitting back to think about how it all works will just look slow-witted. In most classrooms, and among children where academic achievement is everything, the ability to do maths fast is tied to self-esteem. 'Brainy' children jealously guard their position as high-achievers, sometimes because it is the only source of self-esteem they have.

Consequently, many thoughtful children turn their backs on maths, emotionally speaking, because their approach and their needs are seen as inappropriate to maths, but that is nonsense. You can come at it from different angles and be good at it in different ways. That is not the same as saying that anyone can be good at maths or that anyone can reach any level. I like to compare it to athletics: every class will have a fastest runner and a slowest, but we can teach each of them to become a better runner, and with the right encouragement, some of the slower runners might still enjoy it the most.

# What This Book Does

This book offers ways to make maths interesting, enjoyable and challenging. It does this in five main ways.

There are *stories or facts to interest teachers*. This is because the first stage in making maths interesting and enjoyable is for the teachers to feel that way about what they are teaching. Remember that by showing your interest you are modelling that interested attitude to your class. The stories and facts are not all intended for use in the classroom but some teachers may find ways to exploit them.

There are also ideas for how to *make maths fun*. Children can't have fun while learning all the time at school, but this truism is sometimes put forward by killjoys who argue that fun is educationally worthless. I hope you don't think that and won't be swayed by those who do. Most teachers try to scatter pieces of fun into their weekly timetable, and they will get some help here in putting it *into* the maths, not having fun *after* all the maths is done.

There are ways of *introducing new topics* in most chapters. The approach is to ask ourselves why this topic is worth studying, who first developed the ideas and why. This links up to what I say above about a sense of purpose.

But this approach is not limited to introductions. It also works as a way of *opening up familiar topics*, both to a teacher who has had them on the curriculum for years, and to children who are revisiting an area studied to a certain depth in previous classes.

The book also tries to have an effect on the way the lesson is conducted. There are descriptions and explanations of *skills and habits* that make the classroom a place of active enquiry, not passive instruction.

I should make it clear at this early stage that this book doesn't attempt to show anyone how to teach the whole curriculum, or How to Teach, full stop. It is intended as a technique (or set of techniques) to make certain things happen in a classroom. There are other things that need to happen in the classroom which are not the subject of this book. What is more, there are various ways of getting the results we are after. When educational ideas get a bad name, it is often because grandiose claims have been made for what they achieve and how superior they are. The way of teaching I describe need not displace all other ways of teaching – just the ones that you have decided don't work as well.

## How is the Book Meant to Be Used?

The chapters mostly have three parts. The first part is for the teacher to read, mainly for pleasure, but also to get a perspective on a particular topic. Sometimes the discussion is directly about teaching and in others I share my enthusiasm for the philosophy and history of mathematics in the hope that some of it will be contagious.

The second part of each chapter, called *Things to Do*, offers ideas for what to do in the classroom. I have highlighted the key instructions for the teacher like this:

**Do** Take seven boxes of twelve pens (or some similar sets of items) and show them to the class.

**Say** Who can work out how many pens there are altogether? You can open the boxes if you want!

I have kept these to a minimum so that a teacher preparing for a lesson can scan the page and focus on the main actions and instructions. It is unusual to tell teachers directly what to say, but the choice of language is really important, so I have tried to take away the difficulty of choosing the exact wording of instructions. I have had plenty of time to make these instructions as simple and clear as I can, hopefully saving you the trouble.

If the activities strike you as minimal, that is because each one is designed to generate discussion, prediction and a bit of healthy confusion. Your job as teacher is to facilitate the class's journey through all of this. You are not expected to concern yourself too much over what the children will conclude in their discussion, only that the discussion is focused and constructive – not meandering and tangential. As far as possible, I have tried to indicate the likely areas their conversation will cover. But I hope there will be plenty of surprises for you too!

The third part is a *Thing to Say* or a *Key Word*. These are practical tips on what to say to a class to create the right learning atmosphere, or words to describe important aspects of this style of teaching. Some of these key words – such as 'resilience' or 'discovery' – will be familiar to you.

I hope that reading all of each chapter in order will be satisfying, but also that you will find it enjoyable to dip in and out of, trying out different lessons or just thinking about them. The reader can read all of every chapter, in the order they appear, but that is not what I have imagined you doing. You can easily skip to the chapters that sound most appealing or read through the Things to Do for lesson ideas.

At the beginning of the book are four or five chapters with very open topics, suiting a variety of ages and levels, without any predetermined content goal. The intention is for the discussion to stray across the maths curriculum. Later chapters pick out a few big curriculum topics.

The book deals with maths from its most basic levels upwards. It touches on more abstract, theoretical areas, such as different bases and irrational numbers, but I have deliberately avoided the mainstays of the secondary curriculum. That is because I think that primary-level mathematics (whether it is taught in primary school or later) is the key to everything: if they fall in love at that age, they will be set up for life.

# What Difference Will the Book Make to the Children?

The purpose of many of the activities in this book is to make children feel stuck. I am trying to get them to feel comfortable with the 'stuck' feeling as a temporary but necessary stage on the way to success. I regard it as necessary because I want children to learn how to work their way out of it. That is why when children are stuck, it is important to wait with them while they detach their minds from the snag they have got caught on.

We need to be prepared to show real patience. Because if we are impatient about the learning process – as learners, parents or teachers – the process will be incomplete and we end up with half-learned concepts, which we can't build on.

So, if you are going to use this book, what you should be looking for is moments where a child is fascinated, frustrated or foxed. Instead of trying to end or avoid this state by steering children to an answer, you should be trying to exploit it, by getting the child to reveal what is bugging her or asking her to suggest solutions and test them out. In this way, we are helping children to develop strategies to articulate and solve problems.

The benefits overlap and interconnect, but here are some viewpoints.

A sense of *ownership of the learning process* is important. Children want to understand what they are doing and why they are doing it. If there is a lesson objective on the board, it needs to be an objective that is meaningful to the children, not just 'what we are learning today'. Then the work – and its completion – will be intrinsically rewarding; that is, worth doing for its own sake. This book is always trying to grant the class this ownership by provoking their initial interest and training the teacher to capitalise on it.

Children are encouraged to *actively make sense of content* – they ask questions, reflect, experiment and try to generalise. They link what they encounter in the classroom with what they experience outside of it and elsewhere in the curriculum. This also relates to the comprehension and retention of new information. They will understand and remember what they find out in the lesson if they are allowed to piece it together in a way that makes sense to them.

*Engagement with learning itself* is a battle for teachers in some schools. The very idea of what school represents is a barrier with certain youngsters. So, it needs to be demonstrated to the children that they are answering questions that they themselves are moved to ask. The children's self-image as learners and members of a learning community can be enhanced in this way.

The fourth effect that I see in children – although it only develops over time – is *systematic thinking*. I will explain in more detail what I mean by this below, but one thing that fascinates me is that this habit doesn't seem to be linked to intellectual ability or social background. It really is quite an independent skill. Anyway, systematic thinking means going through something methodically as a strategy to get to the best final answer.

Here are some examples of when we think systematically:

* Doing a Sudoku – eliminating possibilities until we solve the puzzle.
* Troubleshooting a DVD or printer by checking plugs, connections, paper, settings and so on.
* Deciding which train to book when looking at a timetable.
* Dividing the children in a class into working groups.
* Trying to find your keys – assuming, of course, that you don't just pace up and down your flat, swearing and looking where you've already looked, just as carelessly as you looked last time.

What do these processes have in common? Well, in each case we:

* Have clear aims or criteria for success (e.g. printing the document, having balanced groups that work together).
* Check things to get information (e.g. Is it plugged in? Are they in my coat pocket?).
* Need to make these checks in the most efficient order (e.g. look in my coat pocket first because that is where the keys should be,

work out what time I need to be at my destination before I look at the train times).

* Should review the sequence and the strategy in the light of the information we are getting (e.g. my total is not divisible by four so if I divide the kids into fours, I will need two groups of five so … ).

These last two points are the ones that really need to be *taught*. Children thirst for the right answer. There are thirty other people in the room who will get the answer first if they don't. So they make hopeful stabs, trying to get lucky – and not taking time to check facts in a logical order. If their first guess doesn't hit the jackpot they just 'roll again' and throw another answer at the teacher, rather than revise their strategy to get 'warmer' by eliminating possibilities.

Impatient adults reinforce this tendency, trying to steer the child from wrong to right as directly as possible, but, in doing so, failing to steer the child from a wrong to a right *way of getting there*.

# What Does the Book Want You to Do?

Supposing you are an adult who wants to steer a child to a right way of getting there – how do you do that? This is where the book can help. First of all, it gives the children things to think about that require systematic thinking. It also gives you tips on how to recognise systematic thinking, so that you can praise it and instil it further.

Most importantly, though, it tells you about *questioning*. The art of questioning is what can make the change. At every stage in a child's enquiry, you are on hand to direct them back to the right question. This is what we do to children when they have lost something and we say, 'Where did you last have it?' What we are doing is giving them the first logical starting point. After a while, they will adopt this question and ask it themselves without your intervention. (You may be thinking that it is an irritating question, but that is because you have adopted it yourself and already thought of it!)

In an exam, for example, you want children to ask themselves questions like this:

- What is this question asking me?

- What information has it given me?

- Do I need all this information?

- What mathematical operation do I need? What tells me this?

- What kind of answer am I expecting? Whole number? Three-digit number? Even number?

So, in the first place you, the teacher, need to pose these questions to the class. Once they are used to you doing this, you can start to say, 'What questions am I going to ask you about this problem?' And before long you can say, 'What questions should you ask?' That middle stage is all-important, though – that is the bit where you are showing them how to do it.

The kind of teaching that I have in mind is more about listening and thinking than telling and doing. If you are one of life's exhaustive planners, I take my hat off to you. But when you plan an enquiry, you should spend your time thinking hard about the whole topic and how it can be explored and understood, rather than trying to determine in advance what will happen on the day. If that takes you out of your comfort zone, well … take it one step at a time. Discomfort is often a precursor to improvement (which I suppose is a posh way of saying no pain, no gain).

To use this book effectively you need to take the role of facilitator, not instructor. The lesson needs to be powered by the children's curiosity, not an externally imposed target. At certain decisive points you will direct the class's attention towards something – but not decide it for them. At other points you may need to intervene and demonstrate solutions – but not until the class has perceived the need for them.

I don't like to make hard and fast rules and this book is not written in that spirit. It is about what you *can* do, not what you *must* do. These guidelines are prescriptive only in that I (and my colleagues) believe them to be the best way to achieve our aims, and they are tried and tested. When I say tried, I mean that I have also tried *not* following these guidelines and it just didn't work!

So, here's how to do the Things to Do bits of this book.

First of all, push all the tables back and have the children sit in a circle on chairs. All on the carpet is OK, but not as good, because the children will tend to speak to the teacher, not each other. This stage is *group-forming.* If they have all just bowled in from the playground, or are divided into tables by ability, are they going to behave as co-enquirers? Probably not. A two or three minute game (something with numbers where you need concentration not computation) is a good way to bond them.

Second, you present the children with some sort of *stimulus.* There is one of these in each chapter. It could be a story, a curious fact, a challenge – anything where the class can see that something needs to be done. Ideally, you will tell them very little about what they have to do. Get them to adopt an active, enquiring role by holding back on your own speaking time.

A good stimulus can be realistic but it can also be highly fictional. Whichever way you go, the typical 'Sami wants to put his collection of seventy-six marbles into eight jars' question won't do. The mathematical problem needs to be seen to arise from a natural desire: why put marbles in jars? Why are there eight? Give some natural context – a clear or believable situation. Give Sami a marble storage issue to deal with first, to which the jars are a real solution, and the limited number of them is explained.

Third, you need to settle on a question for your enquiry. In perfect cases, it will emerge from the class, but not every stimulus has this result. So, set the children a challenge that addresses their understanding of the stimulus. Build up to it. Don't start with 'Find the prime numbers up to 100' – it's not motivating. Before that, we need to get children into the idea of spotting prime numbers, and then 'lift the lid' on the puzzle by asking, 'What is the next prime number?' or 'Is there a way to find them all?' With all of the activities I have given carefully chosen questions to use. With many, I have shown how the questions can progress, gradually luring the children towards the big question at the heart of the matter.

As you read, I hope you will start to see how the stimuli and questions are alike across the book, and be able to make up your own.

Fourth, you elicit as many *strategies* and possible solutions as you can. By a strategy I mean a way of finding an answer. To begin with you might get discouraging conversations like this:

| | |
|---|---|
| **Teacher** | What's the best way to find the answer? |
| **Pupil** | Miss, you have to divide it by two. |
| **Teacher** | Why do you have to divide it by two? |
| **Pupil** | Because that makes fourteen. |
| **Teacher** | And why is that good? |
| **Pupil** | Because that's the answer. |

Why is this discouraging? Because the child is not revealing his strategy, so there is little value to our discussion. You can't blame the pupil here. Or the teacher, come to that. Or at least I hope not, because this is pretty much an exact transcript of exchanges I have had a number of times. The teacher needs to say something like this:

> Maybe you're right. But we need to check that we are doing the right thing. So, who agrees or disagrees that we should divide by two? And why? Can you persuade the whole class?

Notice that the teacher avoids confirming the answer, which is usually best if you want to keep the class's attention on the problem for a bit longer. Admittedly, refusing to confirm the answer can eventually become tedious or pedantic, and you will lose their good faith by being stubborn. But even then you can still put the emphasis on the strategy rather than the answer:

> That's right, fourteen is the answer. Well done. But we are trying to find out why we divide by two. Because some people might not see why you have to do that. And another time, you might not see it! So, who can say why we have to divide, and not take away … or multiply? And why two? Why not … twenty-eight?

You could try awarding points:

- ❖ Two points for explaining a strategy (even if it doesn't work).
- ❖ One point for agreeing/disagreeing with a strategy and saying why.
- ❖ Lose one point for saying the answer before the teacher asks for it.

The point you want to get across here is that you want people to pipe up with whatever idea they have, and not to worry if it turns out not to work. The discussion is about *how* to approach the problem, so one person insisting that they have the answer cuts the discussion short.

You will need to praise all sincere responses, not reject anything straight away. If a suggestion is self-defeating or incoherent, you might invite the suggester to consider the facts that make it so, as this may help to save them from the ridicule of classmates, but try not to dismiss it yourself.

In an ideal world, the class would choose which strategy to test first and then work through all the others. However, in my experience that can be a bad idea; you will end up going down blind alleys and boring everyone, including yourself. It is usually best if the teacher decides which strategy to test – or is prepared to do so. You may select a strategy you know is flawed just because it seems to have a lot of backing from the kids – of course, you have to test ones that don't work as well as ones that do.

The great thing about this way of teaching is that you yourself don't need to know all the answers. In fact, it is often easier to help the children if you don't. Because one thing they really need is a model of enquiring behaviour. If you want children to be curious, committed, engaged and to persevere with difficult things, they need to see you doing it, not hear you telling them earnestly that it is a good idea. That means you have to put yourself in their shoes and struggle a bit.

Teachers are sometimes resistant to going the whole hog and moving the furniture so that the children can sit in a circle. And it might seem like common sense to have children working in small groups, but then they tend to settle too easily for a weak answer or arrive at a solution without needing to justify it.

Also, with the full circle it is easier to take the role of facilitator: interested in the process but not fixing its outcome. Remember that you are controlling the *process* completely – by selecting speakers, asking for more information, going back to previous speakers and so on. But you are not controlling the *content*. They say what they think, and what you are doing is helping them to develop that. By holding back and not confirming any answer as correct, you pull all the children into the process, since whenever someone puts forward an idea, each child has to understand for herself what it is and whether to accept it.

What's more, the class will feel a sense of excitement when they are working together with one set of materials – it is possible to get an atmosphere that is balanced between competitive and collaborative. This is because all the children, regardless of their mathematical ability, understand what is going on the whole time, and feel capable of contributing or commenting. The individuals who shine in this set-up are rarely the exact same ones who score the highest marks when given a page full of sums. So, it gives a boost to a few who don't look forward to maths lessons and can be sobering to those who think maths is just about scoring marks on a test.

# Does This Book Work?

Good teaching is always the ultimate reason for 'hotspots' of good learning. Yes, we could argue forever about how to define 'good' teaching, and it is very difficult. But we don't necessarily need to define it because we know it when we see it (just as we knew it when we got it). Just recently I was talking to a head teacher who told me that she knows within twenty seconds of walking into a classroom whether the teacher is doing a good job. She spends the next twenty minutes making notes on exactly how and why this particular teacher is getting it right or wrong, and it differs from person to person.

What makes me a good teacher, or a bad one, may be different from what makes you good, or bad. The only thing that all good teachers have in common is that their children learn. But it goes further than that because pupils of the very best teachers don't stop with what they have been taught. They continue to learn over time, independently of the school's direct influence, because they were not just given facts to remember and things to do, but a frame of mind – one of continuous learning. That is certainly what I got from a few special individuals in my youth.

And I'm still learning now. All my work is 'work in progress'. I'm putting it out there now for two kinds of people: teachers looking for ways to get their more reluctant pupils into maths, and people who liked school generally but not maths.

The evidence I have is anecdotal. Feedback from head teachers is very often positive. They want to instil a risk-taking, creative, exploratory attitude in all their classrooms. They want all their children to have high self-esteem and to believe they can improve at maths. But it's hard. It's also difficult to reconcile with the barrage of targets, levels, directives and schemes through which a teacher has to pick her way.

Teachers are sometimes excited by my ideas, but sometimes they are spooked or even threatened. There are those, too, who feel it is safe to ignore me altogether. In my view, the doubters are too wedded to the maths-as-hard-labour model or apprehensive about being exposed as weak or ignorant. Of course, I *would* say that. So make up your own mind. But anyone who wants to dismiss what I say should have a better way of achieving *my aims*, because they are important ones. There may indeed be better ways than mine of engaging children with maths and turning that engagement into achievement. That's fine. If someone else's ideas are better, I'll drop mine.

One assumption I have made is that the teacher can get the class's attention and manage behaviour to positive levels. I am as aware as anyone that those conditions are not always in place. I do know, however, that the material and techniques in this book can help win over a class as part of an overall strategy for both ruling and entertaining the young.

# This and That Make Ten

One day, during a discussion with one particularly curious class, we got on to the question of which of the mathematical operations (add, subtract, multiply and divide) we actually need. Most of the class felt that you don't actually need multiplication, because you could just add over and over again, so instead of 5 x 4 you could just do 4 + 4 + 4 + 4 +4. The only reason we do multiplication is that it is quicker, they decided. One child went on to say that you don't even need adding. You could just count. So, if you want to know how many children there are in the school, instead of adding the numbers in each class to make the total you could just count them all one by one. Again, the advantage of adding over counting is that it is quicker.

What I like about this discussion, and the answers I heard, is that it focuses on *what maths is for*. Purpose, in other words. The children understood that human beings developed their maths so they could count more efficiently. This is very easy to forget, and going back to this idea of usefulness is often a great way to open up a new concept in maths (Why do we have percentages? What's the point of learning your times tables?).

One thing that humans began to discover when they got going with numbers is that the methods they used were related in all kinds of ways, that patterns unfurled in unexpected places, and that there were puzzles and problems woven into the very fabric of the number system just waiting to be stumbled upon and solved.

You can see this simply by looking at a grid containing the numbers from 1 to 100: all the different times tables make interesting patterns, and each column and row has a thread running through it. This is a portal to the Numberverse and we want the children to start a journey into it.

| 1  | 2  | 3  | 4  | 5  | 6  | 7  | 8  | 9  | 10  |
|----|----|----|----|----|----|----|----|----|-----|
| 11 | 12 | 13 | 14 | 15 | 16 | 17 | 18 | 19 | 20  |
| 21 | 22 | 23 | 24 | 25 | 26 | 27 | 28 | 29 | 30  |
| 31 | 32 | 33 | 34 | 35 | 36 | 37 | 38 | 39 | 40  |
| 41 | 42 | 43 | 44 | 45 | 46 | 47 | 48 | 49 | 50  |
| 51 | 52 | 53 | 54 | 55 | 56 | 57 | 58 | 59 | 60  |
| 61 | 62 | 63 | 64 | 65 | 66 | 67 | 68 | 69 | 70  |
| 71 | 72 | 73 | 74 | 75 | 76 | 77 | 78 | 79 | 80  |
| 81 | 82 | 83 | 84 | 85 | 86 | 87 | 88 | 89 | 90  |
| 91 | 92 | 93 | 94 | 95 | 96 | 97 | 98 | 99 | 100 |

This first activity is a good way to get them started. What you are looking for is a discussion on the nature of numbers, and the things we do to them. Here are some of the questions that have come out of it:

❖ Is 4 + 6 the same as 6 + 4, or does it just have the same answer? If they are different, does that mean 5 + 5 is different from 5 + 5 if you swap the 5s round?

❖ Is 12 – 2 the same as 2 – 12? And if the answer is no (which it is), why is the *difference* between 2 and 12 the same as the difference between 12 and 2?

❖ Why can you swap an addition round but not a subtraction?

❖ Are there an infinite number of ways to make 10 by subtracting (e.g. 11 – 1, 12 – 2 to infinity)? What does infinite mean?

❖ Is halving an operation?

❖ Are fractions operations or numbers? If they are numbers, how many of them are there between 0 and 1?

❖ Is 1 divided by 2 the same as a half, or are they just equal?

Don't worry if you can't answer all – or any – of these questions. I can't either. And, in my role as a teacher, I don't really need to know if these particular answers exist. What I want in the classroom is for the children to find these kinds of issues intriguing and inviting. Because if they do, they are engaging genuinely with the roots of mathematics itself. One of the best ways to instil this wonder in the group is to display it yourself. So, go on, admit you don't know!

# Things to Do 1

This activity is great with groups that are new to you. You can see what they know, what they half-know and what they've got all upside down. It was first devised for a Year 2 lower maths group with a high percentage of children who spoke English only at school, if at all. Interestingly, it works just as well with older and higher-achieving groups, who motor through the first part and extend it in all kinds of directions.

Take five sheets of paper – A5 or A4 is fine – and write on them like this:

---

| ? | + | ? | = | 10 |

---

**Do** Place the sheets on the floor.

You could put them in the order above or, if you like, jumble them up and let the children take turns to piece together a coherent mathematical sentence – which could turn out to be an enquiry in its own right.

You probably won't need to ask any question at all. Younger children will be bursting to tell you that they 'know the answer – it's 5 + 5!' Almost immediately someone else will point out that there are other answers: 6 + 4, for example.

Let them discuss it for a minute or so then introduce your question:

**Say** How many different ways can you fill in the question marks with numbers?

They discuss it in pairs for a minute and then feed back their answers. It is unlikely that many children will answer your question directly by giving you the number of ways it can be done, but they should, as a group, be able to list the following answers. Record them on the board:

---

0 + 10

1 + 9

2 + 8

3 + 7

4 + 6

5 + 5

---

Watch out for whether they try to do this in a systematic way – for example, by starting with the highest and working down – or haphazardly. If it is the latter, the chances are they will repeat some. Comment on this and encourage any efforts to be systematic. You can say things like: can you see any patterns? What do we do next? Why start here? Have we finished? What have we learned? Why did you do that?

At some point, anchor them (see below) to the original question. Most classes end up with eleven possibilities, from 0 + 10 to 10 + 0, although with some disagreement.

Some children usually maintain that swapping the order of the numbers (from 6 + 4 to 4 + 6) does not give you a new sum, so the correct answer to the question – they claim – is that there are six ways. Get them back into pairs to discuss it if there is disagreement across the group.

Even little ones can now take the enquiry a stage further. Ask if anyone would like to change anything in the number sentence they see on the floor. Give the volunteer a fresh piece of paper and a big pen and ask them to take one sheet out and replace it with something else. One common exchange is to swap the addition sign for subtraction. Whatever they do, you can ask the same question again:

**Say** How many different ways can you fill in the question marks with numbers now?

The beauty of this activity is that within a few minutes of starting it, the class will be deep in discussion on a mathematical point that puzzles them and matches their level of development and understanding. But you don't know what that point is going to be! Although I have pointed

out a couple of the common paths children take, there is always a surprise in there for the teacher.

# Key Word 1: Anchoring

This simple technique is at the very heart of the enquiry method. In
fact, it is so simple that it hardly seems worth calling it a 'technique' at
all. It is this:

---

Repeat the question you started with.

---

The reason we need to remind ourselves to do this is that we can drift
off the topic or people can struggle to make their ideas relevant.
Directing them back to the question we are actually trying to answer is
a very effective and economical way to keep the enquiry focused without spoiling the fun.

For anchoring to work, you have to smile and be enthusiastic, using
positive linking words. For example, if in answering any of the questions above someone says, '5 + 5 + 5 = 15' (which is totally irrelevant),
you may be tempted to point out that this is not helping us to answer
our question. Instead, treat it as an attempt to answer the question,
and say:

---

And … how many ways can you fill in the question marks with
numbers?

---

This is the original question repeated exactly, with the addition of 'And
…' Take care not to say, 'Yes, but …' Keep your tone interested and

neutral. Obviously, there is an element of 'acting' involved here, but it pays off in a number of ways because:

- ❖ The child gets a chance to improve the answer.
- ❖ All the children listening are learning how to be more logical.
- ❖ Other children don't follow this child off track.

Sometimes the child's contribution does turn out to be relevant and interesting once they explain. Even if not, you yourself will often gain an insight into the way they are framing the problem to themselves – which helps you to put them right later on.

# The Real Deal

Numbers are elusive. Take this simple scenario, borrowed almost exactly from the 'Dinosaurs and Plato' section of J. B. Nation's 2003 paper, 'How Aliens Do Math':[1]

---

Imagine two dinosaurs walking through the forest. What does that mean? With no humans there to count them, in what sense are there really two dinosaurs? … Surely we would want to allow those two dinosaurs to be there, whether or not they leave two fossil skeletons for us to count later.

---

One thing he's getting at is that we find it easy to accept the reality of the dinosaurs. What is less clear is whether the number two is real. It is a concept, not a physical object or a word. So if no one is there to think 'two' then does 'two' exist?

Here is another way into the same kind of problem. Have a look at these figures:

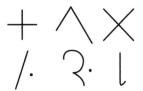

---

1  J. B. Nation, 'How Aliens Do Math' (University of Hawaii, 2003), pp. 2–3. Available at: <http://www.math.hawaii.edu/~jb/four.pdf>.

Think for a while about what they could represent. Why do some have dots?

The red herrings here are the two crosses, because they look like the signs for addition and multiplication. But they are not. Reading from left to right, upper row first, they are the symbols for 10 in Chinese/Japanese, Egyptian hieroglyphics, Roman numerals, Arabic numerals, Hindi (it is similar in many other Indian languages) and Ancient Greek.[2]

What have I written? The same number six times? Six different numbers?

It is worth making a distinction between:

❖ A numeral – which is a mark or symbol, of which there are six above.

❖ A digit – which is part of a number, so European, Arabic and Hindi all have two digits for 10, but written with different numerals.

❖ A number – which is what these written forms refer to, in the same way that dinosaurs are extinct animals that the word 'dinosaur' refers to.

The very deep problem is with that last definition. Not everyone is sure what numbers actually are. In some of the next few chapters, we will find this problem lurking in different places. This book is not about solving those problems for good. It is more about seeing where those problems lie, taking an interest in them and seeing how they can make life difficult for someone mastering a number system for the first time.

---

2  The bottom left character is a real Arabic numeral. So-called 'Arabic' numerals are called – correctly – Indian numerals by the Arabs because that is where they came from.

# Things to Do 2

**Say** It's Lucy's birthday. She is 10 today. For a long time, she has been looking forward to being 10. It sounds grown-up. As she comes down the stairs in the morning, she is wondering what presents she might get. And the more she thinks about it … what she really wants … more than a cake, more than a bike and more than an Xbox … is … a real number 10.

**Say** Can Lucy have a real number 10?

Notice here that the question is in a very simple form. If you want a good discussion, then that is what you need to do. If the question gets vague or wordy or too subtle, it won't spark discussion. Here are some examples of wordings that might spoil the question:

❖ Can Lucy have a real number, like she says, or can't she, because there is no such thing?

❖ What does a real number 10 look like, if it exists?

❖ So … yeah … it's, can she have a 10 really, a real one? That's the question.

❖ Are numbers real though?

❖ So, what's your opinion? You might think she can have a real 10 because there will be one on her birthday card, or she can't because numbers don't actually physically exist, but just tell us your opinion.

When the discussion gets going, a lot of children are inclined to believe that for Lucy's 10 to be real it must have a physical existence. They will also tend to imagine a real number 10 as having the two digits we use to represent it. So it is very useful to have the alternative ways of representing 10 from different languages up your sleeve. Write them on the board but give no clue as to what they might mean. If they get 'warm' I usually let them know that, so they gradually get to the idea that these are the number 10 in various languages. Interestingly, I have done this with classes that knew either the Roman or Arabic versions, but they didn't make the connection in the lesson.

Generally, faced with this evidence, the children will readily accept that none of these 10s is more real than any other.

Look out for which children are agreeing and which are disagreeing, and get them to compare their ideas – you want them to be speaking *to each other* as much as possible, not to you.

**Do**  Write the word 'two' and the numeral '2' on the board.

**Say**  Are these both numbers?

**Say**  'Two.' Which of these did I say? How can you tell?

**Do**  Point to the 'two' and the '2' to show that you might have said either the word or the numeral.

However sure you are that someone is wrong, there is no need to correct them. The value of this activity is in the journey, not the destination.

# Things to Say 1

*'What does that tell us?'*

You are in the middle of a lively conversation and you suddenly have a point to make. You do this by asserting some fact or another. Let's say you are arguing that someone you know is selfish. You say:

---

'He could have given me a lift that time it was raining.'

---

One of the unwritten rules of conversation is that we only say as much as we need to say to make our point in the context where we are speaking. We don't include anything that the listener will already know or naturally assume.

The problem with that is that we leave our listener to infer the implications of what we say, when it sometimes needs to be explicit. As above:

---

'He could have given me a lift that time it was raining.'

'What does that tell us?'

'That he's selfish.'

'Why?'

'Because I needed help and he didn't care.'

---

If this was a social conversation, I would leave it there to avoid being too annoying. But in a lesson, I would add another layer:

---

'So, if someone doesn't care that someone else needs help, that means they're selfish?'

'Yeah.'

---

What I have done is elicited the general principle behind the argument, to add to the specific fact I was already given. If everyone agrees with the general principle they might agree to the argument, but if they don't, then another, deeper wave of conversation will get underway – one that gets to the matter of what selfishness is, and what counts as evidence of it.

# 1, 2, 3, Lots

The story I tell the children for this session is basically true, as far as I know, but I have combined versions from a couple of sources into something that helps the children focus on the key point.[1] This is how I present it to a class:

A university professor went on a journey to Brazil. To get to the place where she wanted to go, she first flew to São Paulo, which is a huge city. Then she flew to another city on a smaller plane. From there she took a bus for two days to a small town, then another bus that went on mud roads and got stuck when it rained. It dropped her at a tiny village by the huge and mighty Amazon river. After two more days, a boat came past that could take her up the river into the Amazon rainforest.

She was going to meet the people from a tribe that she had heard about. They lived in huts and for food they hunted or ate fruit. They didn't use money, had no schools or shops and only a few possessions. But none of this explains the professor's journey. The reason she went so far to meet this tribe was that someone had told her that they only had three numbers.

So the professor started talking to them and found that when people from this tribe counted, they would say 1, 2, 3 and then 'lots'. So, if you held up four fingers they would say 'lots'. And they would say 'lots' instead of 5, 6 and any higher number they were shown.

1  See Daniel Tammet, *Thinking in Numbers: How Maths Illuminates Our Lives* (London: Hodder & Stoughton, 2012) and Alex Bellos, *Alex's Adventures in Numberland* (London: Bloomsbury, 2011).

This is not an isolated case. There are numerous accounts by anthropologists of peoples whose language contains only a few, very low number words. There are various questions to explore.

First of all, we wonder how they deal with situations where there are more than three of something: don't they ever want to know how many *exactly*? In his account, Alex Bellos gives the example of children: if you have more than three children, how do you express it? The answer given by the professor is that children are not seen as belonging to any one person or couple, so the question doesn't make sense.

One more theoretical question in this story is whether the tribe have no *concept* of 4, 5, 6 and so on, or whether they just lack a *word* for them. It is interesting to compare humans with animals here (although I don't usually introduce this idea to the children) because many species of animals can sense whether they are outnumbered when about to attack other animals. For example, they can assess that there are seven of us and five of them and 7:5 is not a good ratio, so they don't attack; whereas if the ratio is 9:5 they are up for a fight. Presumably they don't actually count their potential opponents, but they are sensitive to differences of this kind.

Does this mean that animals have a concept of numbers? I don't know, but I do think there is an interesting discussion to be had about what a 'concept of numbers' actually is, among people who lack the words.

Another interesting area, but one more suitable for children to explore, is whether the tribe could represent higher numbers using the three numbers they have. Children may say the tribe can make 31 with 3 and 1, but that is problematic: what would 31 *mean* without ways of representing 4, 5 and so on, right up to 29 and 30? When, however, children start to say – as they often do – that the tribe can represent 5 with 3, 2 or 3 + 2 we are into interesting territory. If I represent 5 with 3 and 2, do I have a concept of 5? (This is also getting close to the idea of counting in base 3 – see 'Base Jumping' for that path.)

The other area that is touched on here is why we need numbers at all. I can see why we need concepts like more and less, too much and not enough, but how often do we need more exact measurements?

I remember a tour guide in Australia telling a story that might explain why numerical accuracy is unimportant in hunter-gatherer societies. He was saying that when berries appeared on a certain tree, the Aboriginal people would know it was time to head down to the sea because certain shellfish which were good to eat would have appeared in the water by the time they got there. This example helps me to see how accurate judgements could be made without needing any numbers at all; things are just measured against each other.

The more I think about it, the more I feel that precise numbers are only needed for quite complex or specialised tasks, and we can do a surprising amount without them. For example, driving to my home town of Exeter, I don't need to know if it is 56 miles or 72 miles further to go. I want to know if it is about half an hour to drive, about an hour or more. And, in most situations, if I am given a precise figure I instantly convert it to a more approximate one (e.g. 'That's about half an hour' or 'That's about fifteen quid' or 'That's double what I have now'). Maybe we are more like the Aborigines than we realise.

# Things to Do 3

Start your lesson by telling the story of the professor and then try one of the questions below. You are looking to explore issues about the nature of numbers. When one child's comment seems to do that, put their point of view to the whole class to discuss in pairs. Remarks that speculate about how the tribe in question actually live (e.g. 'They couldn't do that because they wouldn't have a leader') are probably less fruitful. Anchor them back to the question if that happens.

**Say** Do you think the tribe needed more than three numbers?

*Or*

> If the people in the tribe say 1, 2, 3 and then 'lots' for everything else, how many numbers do you think they have?

*Or*

> Would we be OK with just 1, 2, 3 and 'lots'?

Children often feel that the main need for numbers is for school (which is a bit circular – we learn maths at school so that we can do maths at school?!) or for buying things.

Some younger ones think that children cannot get older if there are no numbers after 3. It is a common misconception that for every word (or number) there must be a thing, and for every thing there must be a word. Can you have one without the other?

If one child says that they can make higher numbers by putting the smaller ones together, get them to show the class how: hold up, say, seven fingers and they will say it can be represented as 3, 3, 1 or 3 + 3 + 1. Explore this and see if anyone else can build on this or critique it.

# Key Word 2: Ownership

Take a look at children in the nursery. They pick something up, play a while, explore, investigate, exhaust their interest and then move on. The next object may detain them for just a few seconds or for much longer, but their interest rarely flags because they are allowed to choose what they explore and to what depth.

As teachers or parents, we can capitalise on this fascination that children have for their environment. Too often, though, we interfere with this process too much and lead the child by the nose towards something we want them to learn, and the child's interest evaporates. It is because we are too focused on our educational agenda: we want the child to progress, and we have milestones in mind that will reassure us it is happening.

So, basically, don't give kids answers to questions they don't have. Stimulate them to ask questions first. Sometimes you get questions back that you weren't expecting and can't answer (I remember two: what is gravity made of? Is there a number that is not divisible by anything at all?). But that can help you to refresh your own interest in the subject.

My thinking on this was influenced by a recommended way of forming a bond with a young child you don't know, one which is suggested to adoptive parents. The initial technique is to 'mirror' them: hold your hand up when they hold up theirs. Rock from side to side if they do. Once the child becomes interested in your actions you can go from mere mirroring to extending their actions: for example, you hold your hand up and wave. Now *they* start mirroring *you*. This seems to work better than going straight into waving the first time that you kneel down with the child.

We can use this same principle in educating older children. First, get the children to explore and play with the new idea. Reflect their enthusiasm back to them, with only minimal attempts to direct it. Find out what intrigues them: what are their observations and questions? Respond to their observations and help them to answer their own questions.

For example: show children a globe. They might say:

---

If you hold it this way it's nearly all blue.

Why is it tilted?

What do the colours mean?

Is it a map?

Is a flat map easier to use?

How do you find something on here?

---

All these get to the heart of what this piece of apparatus can show us, and what it is for. The answers to these questions will be remembered and reused much more effectively than if you start by saying: 'This is a globe. It's round because the Earth is round. Now look at this side. You can see 70% of the Earth is covered by sea …'

As children mature, they need to be taught in more disciplined ways, plan their learning and become more aware of externally imposed targets. But that is no reason why they can't still own their learning in the same way that the curious infant piling up blocks does.

# Two Square Thoughts

In order to help children get better at reasoning we need to evaluate their efforts. It is important not to judge the reasoning just by the answer, but to observe the strategy the children are employing – or lack of one. In this chapter, I'm going to look at some ways of evaluating children's reasoning, and how to respond to good reasoning, and weak reasoning, in a way that may help them. The task selected has more than one possible answer. However, some responses are still better than others, and I hope to explain why.

When you are trying to encourage more of the strong reasoning, and discourage the weak, you don't want to leave people deflated or spoil the atmosphere in the room. I recommend that when the children are thinking and talking aloud, you focus almost completely on positive reinforcement.

The term 'positive reinforcement' comes from the 1930s behaviourist school of psychology. Their founder, B. F. Skinner, and his followers believed that almost any behaviour could be produced in almost any human or animal by using positive and negative reinforcement – in other words, reward and punishment. If you are trying to train a pigeon to do a somersault, you give it a nasty electric shock when it stands still (that's negative reinforcement) and a food pellet if it jumps in roughly the right direction (that's the positive). The closer its behaviour gets to what you want, the more food you give it and the fewer shocks.

And, to a certain extent, this is how schools are run. We don't usually administer electric shocks (however popular this idea may be with certain members of staff) or throw morsels of food (however reasonable this might seem to certain members of the student body), but we do think hard about how our own behaviour and decisions encourage and discourage the behaviour we want to see.

The reason I would urge teachers to keep to the positive side while doing this style of teaching is that we are trying to deepen confidence and encourage risk. That can't happen if children regret volunteering their ideas. So, if someone makes a point that is obviously weak, remember that you probably don't need to point it out. Often, they will quietly realise it as the discussion moves along or if others say they disagree. I've lost count of the number of times children have put their hands up ten minutes further into a discussion and said, 'I disagree with myself now!'

How do we positively reinforce good reasoning? First of all, don't say the answer is good *because it is right*. In fact, it is very important that you praise some answers that are not, ultimately, right, but do show good thinking. Reiterate the point made and the reason given for all the class to hear – sticking as closely as you can to the child's own words. Here are some of the phrases of encouragement that I use, in case that helps:

---

Well done, that was very clear.

I understand your reason.

That's interesting. What does everyone else think?

Well done for noticing that your idea is a bit like Jack's. In what way is it the same?

I can tell you've been listening because you've linked your idea to other people's.

---

On some occasions you do need to confirm the truth of an answer. I try not to over-reward anyone who gave that answer, especially if that person is seen as a high achiever by their peers. I will say something like, 'Actually, scientists say we *do* have electricity in our brains,' addressed to the whole class, rather than, 'That's right, Wendy, we *do* have electricity in our brains', addressed to a beaming Wendy, top of the class yet again.

So, let's put it all in the context of a lesson.

# Things to Do 4

**Do** Write this on the board, with the spaces between each 2 roughly as shown.[1]

| 2 | 2 |
|---|---|
| 2 | 2 |

**Say** How many numbers are on the board?

Why do you say that?

Are there any other answers?

Before you read some of the children's answers, think for a moment about what your answer would be.

Here are some examples I heard, all in one lesson:

---

4 numbers – because there are four 2s.

1 number – because it is the same number written four times.

8 numbers – because if you add the four 2s you get 8.

---

1   This activity also appears in Peter Worley (ed.), *The Philosophy Shop* (Carmarthen: Independent Thinking Press, 2012) and Peter Worley and Andrew Day, *Thoughtings: Puzzles, Problems and Paradoxes in Poetry to Think With* (Carmarthen: Independent Thinking Press, 2012).

5 numbers – the four 2s plus the 8 that you get from adding them together.

2 numbers – 22 and 22.

1 number – this is a four-digit number written in two rows.

1 number – because it's a fraction with the bar missing.

12 numbers – four 2s, plus four 5s, because the 2s are 5s upside down and in reverse, and upside down 7s with tails.

0 numbers – because these are just representations of a number, not the number itself.

___

Obviously, in an exercise like this we won't now choose one of these answers and say it's right. But that is not the same as saying that they are all equally right, because they are not. What we want to look into at this point is the rightness and wrongness of each of these answers. Some have a lot of one and not much of the other.

Taking them one by one, here are my evaluations. I don't intend any-one to take them as gospel, and I have discussed them with colleagues who have a slightly different take on some of them. The purpose of including my thoughts in detail is to show that we can have a middle ground between, 'You see? That one is correct,' at one extreme and, 'All your answers are lovely' at the other. Read as many of these as you need to, just to get the idea.

___

4 numbers – because there are four 2s.

___

This would be many people's first answer and represents common sense.

___

1 number – because it is the same number written four times.

___

This is a more reflective answer. It is less obvious than the previous one, but clearly true in one sense. It could be seen as a better answer because it requires more thought, although that doesn't make it more true.

---

8 numbers – because if you add the four 2s you get 8.

---

This avoids the obvious, but it doesn't make much sense as an answer to the question, 'How many numbers are there on the board?' The big problem is that the child has confused the number and the number of numbers! Because if you add the four 2s together you do get 8, but that is one number not eight numbers. Also, I'm not sure that the number 8 is on the board just because there are some numbers that add up to eight.

---

5 numbers – the four 2s plus the 8 that you get from adding them together.

---

I think this answer is a bit more coherent than the last one. I disagree with the adding thing, but I don't think it involves an obvious error or contradiction.

---

2 numbers – 22 and 22.

---

This is someone trying to think outside the box. It opens up the chance for discussion about why we leave spaces between numbers – and it is the same reason that we leave spaces between words: to show where they begin and end.

---

1 number – this is a four-digit number written in two rows.

---

Here, it is worth finding out if the child can think through the implications of this. I mean, why *can't* I write a four-digit number in two rows? The answer is convention. There would be nothing to stop us having a number system where we write the digits in rows – as long as we understood the place value of each place, we could read the number. The thing is, of course, that the person reading the number needs to know the convention being used. If you get any discussion of these issues from the children then the answer is a good one in the sense that it is thought-provoking.

---

1 number – because it's a fraction with the bar missing.

---

Fairly weak, but you could compare this to a word example. If I write 'bead' on the board, does it have to be the word 'bead' or could it be 'beard' with the 'r' missing? Or is it actually 'bead' and that is that? And does it matter if the writer missed out the bar on purpose or by accident? Does that affect what the number is?

---

12 numbers – four 2s, plus four 5s, because the 2s are 5s upside down and in reverse, and upside down 7s with tails.

---

This is far-fetched. Basically, we are being asked to accept that the figure 2 has other numbers/figures in it. This is quite an irrational attitude, rather like believing that if I say the word 'human' I have also said the word 'hum'. I think it is worth getting the child to make their logic explicit. In other words, does every number that has a straight line in it contain a 1?

---

0 numbers – because these are just representations of a number, not the number itself.

---

This is the most philosophical answer, in some ways. That doesn't make it right, or even the best, but it does show an ability to think beneath the surface of things, so understanding and appreciating this idea is good for children. It only rarely comes up from the children, so I introduce it myself. One way of doing this is to say, 'A child in another class told me there were no numbers. What do you think his reason was?' By using a pretend third party I am not introducing the idea as mine, which would kill off the controversy as very few children are comfortable disagreeing with the teacher for long.

Another way to do this is to draw a picture of a flower (or a stick man, whatever) on the board and ask how many flowers there are on the board. You have a pretty good chance of a child pointing out there are no flowers, just a picture of one. You may already be thinking of René Magritte's painting *Ceci n'est pas une pipe* (This is not a pipe), which is a painting of a pipe, with the words of the title written underneath it to remind the viewer that they are looking at a *picture* of a pipe, not a pipe. And you could bring that in too.

**Do** Extend this discussion by showing:

    a) Two shapes – either identical or different in some way.

    b) Two coloured patches – both the same colour, or different colours, or two shades of the same colour.

**Say** How many shapes/colours are there here?

Is your answer the same as for the numbers? Why/Why not?

See how their answers compare to the ones they gave earlier. If they think differently, what has changed?

Notice that I have compared numbers to words and pictures more than once in this task. Whenever we are thinking about what a number really is, these comparisons can be useful. Numbers are like words and pictures in some ways, and different in others.

# Things to Say 2

*'So are you saying … ?'*

As well as identifying certain kinds of reasoning as strong, let's look a little more at how to teach children to reason better.

First, it is important to identify the *assumptions* behind weak reasoning. Get the child to commit to the assumptions that they need for their argument to work. In the '12 numbers' answer above, the child needs to state explicitly that because chopping the curly bit off the figure 2 and turning it upside down makes it a 7, that means the figure 2 *actually* is 7 as well.

---

So are you saying that you can chop a bit off and make it into a 7, which means it is 7?

---

Of course, tone of voice is everything here. You could say these words in such a way that the child immediately retracts what they had just said because you are showing your disagreement. So, use an encouraging tone of voice for all answers, and don't let slip what you think. A lot of teachers who have observed me say that concealing their own opinion would be the hardest part for them in doing what I do. But if you want to see children think their own way out of a problem, you have to reserve judgement. This is not the same as letting go completely. With your questions you can be constantly focusing their attention and getting them to think about the right kind of things.

So, once the questionable assumptions behind an idea have been highlighted, it is up to others in the class to expose the flaws. If they don't, you know that your class is not reasoning strongly as a group, but don't be impatient to prove that to them. Trust that when one child starts to do what you want, your positive reinforcement will be picked up by other pupils, and soon more of them will work it out for themselves.

In the same way, *get children to commit to contradictions or errors* by repeating them. So, you would say:

So are you saying that because the four 2s add up to 8, there are eight numbers on the board?

A lot of the time this will trigger a rethink in the speaker, but not always. And even if not, you have now made it easier for other children to correct the reasoning. And don't only highlight weak reasoning. Because then it will just become a signal that the teacher thinks it is wrong. Regularly reiterate interesting points for the whole class to consider – they can discuss those points in pairs before feeding back their ideas.

One last word of warning: teachers often use the phrase, 'So are you saying … ?', and then continue with something the child didn't say at all. This is often done – consciously or unconsciously – to improve what the child said, or to bring it round to the teacher's current train of thought, which is not our aim. Stick to the child's wording and keep to an innocent, questioning tone as you check what they mean. If you have understood them correctly, they'll nod firmly. If you haven't, they may still say, 'Yes', but you'll know from their body language that you haven't at all!

# Base Jumping

When we count out loud we say: 'One, two, three, four, five, six, seven, eight, nine, ten.' But when we write those numbers down, we put: 1, 2, 3, 4, 5, 6, 7, 8, 9 … *10*. The last of those numbers is written in a different way from the others. Instead of a new symbol (which is what we have for each number up until then), we write one of the symbols we've already used and a zero. This is called counting in base 10. It doesn't occur to us that there is any other way of doing it until it is pointed out.

Why do we do it? We could, for example, use another symbol – let's say ■ – to mean 10, and another – let's say ● – to mean 11, so that the sequence would carry on like this: 7, 8, 9, ■, ● … and we could have a new symbol for every new number, going as high as we like.

Why don't we use that system? Well, you may already be able to see some disadvantages to it. First, we would have to memorise a lot of symbols. And we might hardly ever use some of them. After all, how often would you need to use the symbol for 972? And how would you understand the relationship between numbers if they each had their own unique squiggle? Would you notice that 200 is half of 400 if they weren't written like that but had their own symbols?

One way to illustrate this to children is to get them to use only 1p coins for transactions and calculations (which we will do in the next chapter). They will readily see that introducing 10p coins, although not strictly necessary, is far more practical. That is basically what counting in base 10 is.

By counting in 10s, and putting a 1 in a new column to represent one set of ten things, we can represent numbers efficiently, logically and in a way that allows us to calculate. But why 10? Why don't we count in 8s or 12s?

The answer is probably right in front of you, right now: your hands probably contain ten fingers. Humans learned to count in 10s because we ran out of fingers at ten. Now compare your hands to those of cartoon characters. Many of them have four fingers on each hand – take a look at *The Simpsons*. Why do they have eight fingers? Apparently it looks better! Anyway, if humans had eight fingers, they would probably have counted like this: 1, 2, 3, 4, 5, 6, 7, 10 – where '10' could be pronounced 'eight', or 'ten', or with any other word you like. The point is that the eight-fingered people would not need the symbols 8 or 9, and if they looked at a spider they would write the number of its legs as '10' – although they would mean exactly the same number we mean when we write '8'.

Counting in 10s or 8s is known as using different bases. Have a look at some numbers in base 10 and base 8. (When saying numbers in a base other than 10, we usually pronounce 10 as 'one-zero' rather than 'ten' and 23 as 'two-three' rather than 'twenty-three'.)

| Base 10 (our numbers) | 1 | 8 | 16 | 24 | 64 | 65 | 512 |
|---|---|---|---|---|---|---|---|
| Base 8 | 1 | 10 | 20 | 30 | 100 | 101 | 1,000 |

Although we use strict base 10, there is evidence that way back in the past our counting system was more mixed, or completely different. Just think, we have special words for the numbers eleven and twelve even though they are in the 'teens' and should be one-teen and two-teen. Is this a sign that we once counted in base 12? Even if not, the words 'dozen' and 'gross' (meaning 144, which is 12 x 12) were common until recently because they were used so much in trade.

In French, the counting system kind of stops at sixty – *soixante*. After that they count on up to sixty-ten (*soixante-dix*), four-twenties (*quatre-vingts*) and four-twenties-ten (*quatre-vingts-dix*) before arriving at one hundred (*cent*). Not even the most obstinate Frenchman could claim that this makes much sense. But there is a kind of fossilised logic to it, because it probably remains from the days when people used sixty as a round number rather like we use 100. It is a handy number, dividing into 2, 3, 4, 5, 6, 10, 12, 15 and so on. And there is evidence that several ancient civilisations, such as the Sumerians (back in 3000 BC), did something similar. Not only that, but dividing the day into twelve hours containing sixty minutes is said to date back to ancient times, and base 12 systems.

You might think that counting in anything other than base 10 is hopelessly primitive, but computers would disagree. Computers count in base 2. This means they only use the digits 1 and 0. So when you get to 2 you write 10. Here are the place values in base 10 and base 2 compared:

| Base 10 | 1 | 2 | 4 | 8 | 16 |
|---------|---|----|-----|-------|--------|
| Base 2 | 1 | 10 | 100 | 1,000 | 10,000 |

So, our 19 would come out as 10011. This seems pretty daft to begin with, but long numbers don't bother computers much. The reason they use base 2 is that it uses only two symbols (1 and 0) which can be represented by the on or off positions of a switch. Computers love that.

Seeing as most children – and perhaps most people – find the idea of different bases mind-boggling, it may seem very counterproductive to introduce them to it. But their confusion stems from a misconception about numbers: the idea that '10' somehow *is* the number 10, when it is really only a way of representing a group of ten things in an efficient way. It is not actually the most efficient way possible, unfortunately, as it would be easier to calculate in base 12. But deep down, I don't believe we can really understand the way we *do* count, unless we compare it to the ways we *don't* count.

Going as far as calculating in different bases may well be something for only those children who enjoy working with numbers for the fun of it. But introducing the concept of base 10 as just one of a number of systems can help to dispel some of the problems with place value that can dog many a child's progress.

# Things to Do 5

**Do** Tell the story below.

---

Imagine there are some aliens on another planet, called Octies, who have four fingers on each hand – so eight all together. Because they have eight fingers, they like to count in 8s. They have a different way of writing down numbers. They start the same as us: 1, 2, 3, 4, 5, 6, 7 … but then something very different happens.

For them, the next number after 7 is called 'one-zero'. They write it like this: 10. After one-zero they count on in this way: one-one, one-two, one-three and so on. Here are their numbers written down:

1, 2, 3, 4, 5, 6, 7, 10, 11, 12, 13, 14, 15, 16, 17, 20, 21 …

As you can see, after one-seven, Octies go to two-zero and two-one.

---

Add or drop background details as you prefer. I use aliens for lots of stories because children seem to accept that the situation is imaginary but possible – and that is very important. You don't want objections from children that this situation can't be real. If you do get objections, accept them and simply say, 'Yes, maybe you're right. And *if it was real, then …'*, and repeat the question.[1]

The first paragraph is quite straightforward but the second one has some big concepts to get your head round. Children will slip up by saying 'ten' instead of 'one-zero' when they see the numerals 10 in base 8, and will really struggle to see that 14 in base 8 is the same number as 12 in base 10. They identify the number totally with the way of writing it and can't separate the two. So, tell this part of the story very carefully and go back over it. Remember, the only point in covering this territory is for them to get a glimpse of this concept; it is not for them to successfully calculate in other bases.

Get the children to say the names of the numbers in base 8, first saying them in order and then picking them out randomly one at a time. Expect lots of mistakes, and give more time for practice. Then:

**Say** Can you write the Octies' numbers after 21? How do you check you are getting it right?

What do you think of Octie numbers? What should their way of counting be called?

**Say** How many legs would they say a spider has?

How many fingers would they say we have? What about fingers and toes added together?

**Say** Could you explain to the Octies the difference between the way we count and the way they count?

If the Octies understood your explanation, would they like to switch to our way of counting? Is our way or their way better?

---

1 My colleague Peter Worley coined the term 'iffing' for this technique – see *The If Machine: Philosophical Enquiry in the Classroom* (London: Continuum, 2010), p. 35 for his explanation. It is actually a specific way of anchoring the group to the original question, but a very important and powerful one.

**Say** How do you think twelve-fingered beings might count? What will they need after nine that we don't have?

There are various activities you can do to follow up:

❖ As a group, you can translate numbers in base 10 to numbers in other bases. Let them go as slowly as they need and use counters to make sets.

❖ They could also translate numbers from other bases into base 10, and vice versa. Draw or use counters if it helps.

❖ It is possible to do additions and subtractions in another base. More able children might enjoy experimenting with this.

❖ Show the number system for base 7, then base 6 and keep going down – what is the lowest base and what does it look like?

# Things to Say 3

*'Who understands what he is saying? Can you explain his idea?'*

There are times when one member of a group makes a breakthrough. A child puts up his hand and lets out some new idea that suddenly lights up a whole new path. As you see heads nodding and voices murmuring, 'Aha!', it is easy to perceive a ripple of comprehension running round the room, and feel at last you can move on.

Be careful though. How many children really understand? Some will indeed understand what has been said but miss the wider significance of it. Some will realise that the teacher likes the answer and will pretend, hopefully, that they know what is going on. Others are keeping their heads down, hiding – just thinking, 'Now I *know* I really don't get it.' If you move on straight away, all of them will end up lost before long.

These people must not be left behind. If they are, they will be left behind not only in knowledge but also in self-esteem. The maths classroom will become a place where they don't expect to follow their peers and they will develop other ways to boost their self-esteem or occupy their time.

Here are a few things you can do:

- After the child who has had the insight has finished talking, ask for someone else to explain it so that everyone can hear it again. Whoever volunteers may do a good job, in which case more of the rest of the class will probably start to twig. Or they may struggle, proving that they didn't follow the idea at all. In this case, go back to the original speaker for clarification and/or let another person have a go.
- Get them into pairs. Say, 'Alex says that two-fifths is the same as 2 divided by 5. Is he right?', and let them puzzle it out. Go back to the group discussion.

Quicker, abler children will stay engaged if they feel the class is discussing their idea. They will also improve at explaining their ideas if they get feedback from peers and the chance to reword what they said. It is quite common for these types of children to be socially isolated because they think other children are not up to their level, so the need to explain themselves and the opportunity to do so are good for them.

# A Jump to the Left … and a Step to the Right

Fashions change in education. What one generation sticks to as a tried-and-tested method is rejected by the next one as backward and prejudiced. The generation after that throws out the trendy approach and rediscovers a traditional one. To and fro, to and fro. And the fashions overlap, so that one set of people could be discovering a fad while another set is already caught up in the backlash from it. I sympathise with teachers and parents who are not quite sure of the current tide, and whether it matters.

I remember our teachers telling us sternly that, 'the decimal point does not move', and that if we were ever told by our parents that it did, we should put it out of our minds. The moving decimal point was part of our parents' old trick of multiplying with decimals by (a) counting the digits after the decimal point and then (b) putting the decimal point that number of digits into the answer. For example:

---

46.25 x 7.1

---

They would treat this as 4625 x 71, giving an answer of 328375. Because there are three digits to the right of the decimal point in the original sum, they count three digits into their answer and put a decimal point there:

328.375

The problem, apparently, was that it was a mere trick. And tricks, it was believed, didn't help us understand what was really going on.

I can't quite remember the method we were taught instead. This may be because, at some later point, one of the other kids showed me the 'bad' decimal point trick we were supposed to steer clear of and it made multiplying decimals a lot quicker.

I can see why many people believe that tricks are not worth learning. For a start, calculating is easier these days because we have calculators in our phones to do it for us, so it is surely more important to understand the wider process than to plough through sums that machines can finish in a nanosecond. If learning handy tricks interfered with our conceptual understanding of the maths that would be a real worry – though does it, really? I don't know.

Although I respect the argument against crafty tricks, I'm not too sure why telling children that the decimal point moves is such a sin. It's a bit odd, especially if instead you say that the digits move to the left if you multiply by 10, as I was taught. Is it really an aid to deep understanding? Look, if we multiply 8.3 by 100 then we get 830. Yes, the digits 'move', but they represent utterly different numbers in their new positions – that is the whole point of a place value system. And it's the same with the trick of 'adding zeros'. If I have to multiply £120 by 200 it is good to know that because I have three zeros in those two numbers, I will have three zeros in my final answer: £24,000. That is not deep understanding, but it is a nice rule of thumb to keep me on track.

So, as I launch into this discussion of place value, I am beset by doubt! One thing I remember clearly from my own school days is that I found it very hard to accept that 10 was 'one 10 and no units', as I was repeatedly told. The numbers up to 9 weren't complex in that way, and I didn't want 10 to be either. I wanted 10 to carry on being just … 10.

My peculiar denseness on this issue may not be shared by all children, but I do think that when grappling with place value children have to learn to accept that there is something artificial about our numbers – that 27, for example, is already broken down into different groups of 10s and units. This seems to make numbers kind of real, but not real. In fact, one of the most ancient philosophical conundrums is whether numbers are real. If I've got ten fingers, people wondered, then the fingers are real but what about the ten? Are numbers things in our heads or are they actually out there – like the fingers? Or are they out there in a different way from fingers but still real?

The same kind of questions can be asked about the words we use, but then the answers seem simpler, because it is easier to distinguish between the word and the thing. If I say, 'I'm eating an apple,' the apple exists but the word 'apple' is a made-up thing to represent it. If I was speaking French, I'd call it a *pomme* and in Turkish it would be *elma*, and so on. There are two separate things – the physical fruit and the word we happen to use to refer to it. Some philosophers (followers of Plato) would claim that there is a third thing in existence: the *idea* of apple. The idea exists, they say, just as much as the actual apple I happen to be eating. In fact, they believe that *the idea is the only thing that is completely real*, because once I've eaten the apple it's gone. Or I could be having a dream where I am eating an apple, which means it's not real. And even if all the apples and apple trees became extinct then the idea of apple will still be there because ideas can't be eaten or die out.

Is it like that with numbers? I mean, does the word 'four' refer to a thing or a number? And if the number is a thing, independently of what we name it and how we write it, what kind of thing is it? Because numbers aren't like apples or fingers, as I said, or even words.

This may seem far removed from how to do a page of sums! And in some ways it is. But what I'm trying to provide is a glimpse of the uncertainty that surrounds numbers: are they not real because they aren't physical, or are they the most real thing of all precisely because they aren't physical – which makes them eternal and indestructible?

I'm not suggesting that it is necessary to answer these conundrums to learn maths. It is just that if children see the conundrum, however dimly, they can be confused by it. And it is so frustrating to have doubts while thinking we are supposed to understand. Once we know that our doubts are natural and appropriate to the subject, they are less likely to interfere with our progress, even if they aren't completely answered. Get children talking about these issues and you will see that they genuinely do peer into the philosophical depths …

The next set of tasks is my way of teaching practical stuff about place value, but in a way that confronts some of the confusing aspects, so that the class can puzzle over it a bit, the teacher can agree it is puzzling and then all agree that there are ways of making sense of it.

# Things to Do 6

**Do**  Get a whole heap of 1p coins, and about ten 10p coins. Add a 50p coin if you want to do subtraction as well.

Schools often have fake plastic money instead, I know. But see how much more attention you get with the real stuff. You can get 1p coins over the counter from a bank – they come in little clear plastic bags. With £5 worth, you should have plenty.

**Do** Get some real objects that the children might regard as realistic purchases (e.g. food items, magazines, stickers).

Price the items up clearly, with numbers well into double digits

Get some of the children to come to the front and play shop.

They take it in turns to buy things from the shopkeeper. Before long, the counting of the money will be laborious. Once this has become clear to them, ask:

**Say** What do we need?

You will soon get the answer, 'Bigger coins', or something along those lines. So …

**Do** Introduce the 10p coins. Repeat the activity with these. Get them to buy multiple items so that they have to add up.

Make this easy at first so that they can add without 'carrying' (e.g. 21p + 14p). Let them get confident at arranging amounts into 10s and 1s.

Then comes the key moment. You want them to discover that each 10p coin can be broken into 1p coins without losing any value – or gaining any, for that matter. To do this …

**Do** Give them a problem where they have to 'carry'.

An example would be 16 + 9. They can count this out by taking 16p (made up of 10p + 6p) and adding nine x 1p coins. This gives them fifteen 1p coins. When they come to hand the money over, do they convert ten of those 1p coins into a 10p coin or not? It would be much more practical to do that, so watch to see if anyone does. If not, ask:

**Say** Can anyone see a way to speed this up?

Can we use the 10p coins to speed this up?

If all this is revision of concepts they can already work with, then it can serve as good preparation to have a go at subtraction. So:

**Do**  Get the shopkeeper to give change, by giving them all the 1p coins and giving the customers a 50p coin.

They will now have to do things like 50 – 27 or 50 – 18. To do this, they can put together 50p from the smaller coins (e.g. 4 x 10p and 10 x 1p) and then it is easy to take away the cost of the item from that amount, and what is left is the change. Once they are confident with this, they are acting out the process that we want them to perform when calculating subtractions – and dealing practically with the place value. They are seeing that you can't take 8, for example, from 50 until you've broken that 50 down into its constituent parts.

If your class already has additions and subtractions like this under their belt, then next you might be teaching them how to multiply by 10. And you can use much the same idea. This time you will need more of the higher denomination coins though. Again, use the coins as money to buy goods. Say that bananas are 10p and give out 10p coins for the time being. They will quickly establish, if they don't know already, that two bananas will cost 20p, three bananas 30p and so on. Get someone to record it methodically for all to see. Ask someone to point out the pattern in the written form – we are adding zeros.

The next thing to do is flip it round, so instead of 3 x 10 you have 10 x 3. If the children have really got the hang of multiplication, this should be a trivial thing, but don't assume it will be – check. Find something that could be 3p and buy ten of them, something that could be 4p and buy ten of those, and so on. Establish that we have the same pattern of results – we are adding zeros again.

Now take the item that is 3p and say you want 100 of them. Leave the class to work out in pairs or groups how much that would cost. Once some of them have got the answer they can focus on trying to *prove* the answer. Do you need to count out 100 lots of 3p? That would be fairly reliable but slow, though maybe one of the class should set about doing it – just to test that it works while the rest of them move forward.

Soon you will have a second table of answers reading 100 x 2 = 200, 100 x 3 = 300. Every single child in the class needs to have seen that this must be true, and then they should all try to remember that to get the answer you add zeros. You can then move on to multiplying by 20 and 200, 30 and 300. With any luck, they will know from now on that we add zeros, or the digits jump to the left, or that an independent and self-willed decimal point has hopped over some zeroes – whatever story works and/or is part of the school's adopted methods. But they will have seen for themselves, in silver and copper, that it does work, and how.

# Things to Say 4

*'I should have taught you this.'*

I think it would be great if teachers said this, but we don't. One thing we hear instead is, 'You should know this.' It is a fairly natural reaction to the sight of blank faces. We think, 'Come on, I told you only last Friday' or, 'You did this last year.' It certainly is disappointing to discover that something you went over conscientiously and comprehensively – and you checked that they understood it, and they all did it over and over – is now the wispiest of vague memories floating in the ether a few feet above their heads.

Maybe it's not the teacher's fault. Maybe the kids are not serious about learning. Maybe they don't have good memories. Maybe this, maybe that. After all, you did your best, but what goes on inside those skulls is beyond our control at the end of the day (*especially* at the end of the day).

But, in the end, it really comes down to how good a teacher we want to try to be. And if we are trying to be as good as we possibly can be, we have to accept this basic motto: *teaching cannot take place without learning.*

In other words, if they don't know it, I didn't *teach* it. Yes, I may have talked about it, set work and done everything humanly possible to help them learn – but I didn't teach it, because if I did they would have learned it. I don't say this because I think we should be beating ourselves up about it. I say it as a possible way forward – an alternative to shrugging and disbelief. It is about taking *responsibility* for their learning. Ultimately, yes, you want the responsibility to be taken by them, but they will learn that seriousness of attitude from you.

# A Plus Point

Here is a thought that might surprise you. One problem in showing children how to add is that we ourselves might have forgotten how to do it. That may sound odd, but it may well be true. Look at this set of sums and think of the answers:

---

$$3 + 2 \qquad 4 + 5 \qquad 3 + 8 \qquad 2 + 4$$

---

You didn't add them. You didn't have to – because you could simply recall the answers. We all memorised these combinations long ago, so long that we can't remember what it's like not to know them. If you say '7 + 2' to me I automatically think of 9. Just as if you said, 'The United States of …', I would think of America. So, when we educated adults 'add' small numbers together, we are really just reading from an answer key in our heads – not calculating.

One teacher was concerned that some of her class weren't seeing the difference between adding and counting. 'They're getting the answers,' she said, 'but they see it as counting.' She was a person that I'd learned from in the past, so I started asking myself, just what is the difference between true adding and mere counting?

A mother I knew was also bothered by the way her daughter was learning to add in school because she felt that the girl wasn't being taught addition at all. It was just counting. I'll describe it:

---

We're going to do adding. So you throw the dice and see what it says ... 3. OK, so count out three cubes. Now throw the dice again ... 2. OK, so now count out two. OK, now we're going to add the 3 and 2, so move both piles together ... that's right ... and count how many you have now. That's right ... 5.

---

Now, I'm not saying that this should never happen. Or that it's a bad lesson, method or activity in itself. And nothing is easier than to shoot down something you observe when you haven't seen what came before or afterwards. However, I would like to look closely at it. I like the fact that the children physically put the piles together so they see what addition is in a concrete way. But I sympathise with my friend's objection that the rest of it is not really adding. You just put them together and count everything, ignoring the fact that you had previously counted the two piles. To me, counting again from scratch should be a backup, a way of checking.

So there is an unexpected problem here. If counting is not really adding, and what you and I do when we are 'adding' is not really adding ... then, what is adding?!

The answer is a bit elusive, but I think we would all agree that adding is increasing, making something more or bigger. It is also about making two things into one thing, or thinking about them that way.

At this point, you could say, 'Who cares?' Does it really matter if we are 'really' adding as long as we get the answer right? My answer to that leads me back to the lesson described above and to my second problem with it: what is the point of adding? We need to see the point when we do it. So, in this case, I'd prefer we didn't roll dice. I'll admit that it is a way of quickly generating a lot of addition problems randomly – but, let's face it, that is not exactly hard to do. Whereas what is missing is a sense of purpose, a reason why a person would want to go about doing this whole 'adding' thing. Without purpose you just have an empty classroom exercise, not rooted in life experience and therefore unlikely to be applied to it. And if that is the case, then children are already, at the age of 5, learning to dissociate what they learn in the classroom from what they see outside of school.

To get it right, I think we need to create a realistic context for what we are teaching and focus on that. The key thing is to think of a situation where we have a set of things and want to add another set of things to it. The story we weave to set up the activity can be very short, as long as we make it engaging for the class. So here goes …

# Things to Do 7

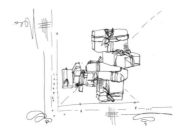

My default choices for getting the attention of children are money, food and – as in this next activity – presents.[1] For the best results, wrap up some presents or print and cut out some pictures of presents.

---

1  If you're in a multicultural setting, it might be useful to know that most families in the UK celebrate birthdays in some way, except Jehovah's Witnesses. Therefore, if you have a child with that background he may not find it so engaging, though he's not forbidden from talking about birthdays. Just don't tell him, 'Imagine it's your birthday …'

**Say** It's Tara's birthday. Her mum tells her that she can open her presents when she gets back from school. She spends most of the day at school thinking about what is waiting for her back home. When she finally walks back through the door, her presents are laid out on the table. Here they are. She counts them. How many are there? That's right, four. But then Tara's mum says that her auntie and granny are coming over. Auntie has a present and so does granny.

**Say** How many presents will Tara get for her birthday altogether?

What you want the children to do next is think in this way:

---

Tara already has … four.

One more from auntie makes five.

Another one more from granny makes six.

---

*Or*

---

Tara already has … four.

Auntie and granny are bringing two more.

That makes five, six.

---

There are two differences between this set-up and the style I criticised. First, we have a meaningful context. Second, we don't count the whole group again from zero but continue from the total of the first group.

To extend and give practice, tell the same story but for Tara's birthday the following year.

**Do** Change the number of presents she starts with, but still have granny and auntie bring two more. Keep going until everyone has got the idea that you count on two. Lower the number so that they are adding 2 to 1, then 2 to 0 (it's easy to get mixed up when adding a number to one that is close to it). If some children are starting to find it easy, tell them to detach from the group and move on. They could:

❖ Think of other stories with adding, using Tara or a character of their own.

❖ Practise writing down what happened: 4 + 1 + 1 = 6 or 4 + 2 = 6. Now they are using mathematical notation they can extend the problem easily, by using bigger numbers or doing several additions at the same time – for example, a whole extra bunch of relatives turning up with presents.

# Things to Say 5

*'Try again.'*

Our problem is that we just don't have enough time. It's not always our fault – it's modern life. And there are pressures on teachers to deliver on measurable goals within time frames. Any intelligent person in this position would be on the lookout at all times for ways of getting most directly to those goals. But what learners need from teachers, above all else, is feedback and encouragement. And these can't be rushed.

When we are trying to guide someone through a new process, such as when they are learning to drive a car, there will be times when they fail at something. The tendency is to jump in with feedback immediately – to *tell* them something. Sometimes this is vital because there will be no point in them making another attempt if they go about it in the exact same way. They need to know why they haven't got it right up till now and what they need to change.

There are other times when you don't need extra or repeated information. What you need is encouragement. Encouragement can mean a lot of things – don't automatically think of some crazed sports-coach type whooping and high-fiving and telling us to reach for the sky. Often the most encouraging thing we can do is show that we aren't worried about the failure, that the pressure to perform has not increased, but decreased if anything.

We can do this with calm, neutral body language and a relaxed, friendly voice. Imagine you've just messed up a three-point turn in your driving lesson. You know all the things you have to do perfectly well; you just messed it up, that's all. Would you prefer the instructor to breathe in hard and say, 'OK, don't worry. What you've got to do is …', and reiterate the procedure, or just smile at you and say, 'OK, try again'?

Some of my students know that when I say, 'Try again', I'm actually telling them they got it wrong – I wouldn't have said it if they had got it right. But what I'm getting them to focus on is the next effort, not the past failure. And in two words.

# Don't Drag the Tables

Much as I love to wave the flag for creativity and independence and all that liberal progressive stuff, I know there are times when people just need to knuckle down and memorise. As learners, we also know that it is sometimes what we need – we're saying, 'Just show me what to do so I can copy you.'

The perfect example in maths is multiplication tables. This is a set of data that needs to be uploaded into the memory, ready to be called upon at any time. So, I'm not against rote learning at all. We just need to do two things: rote learn only the things that *need to be learned that way*, and demonstrate the *need for that data* to the student.

So, what are the things at school that must be learned by rote? Facts and figures which give us a framework to navigate our subject. Casting around a few different subjects, here are some examples:

❖ Dates of major wars and battles in history.

❖ The five pillars of Islam in RE.

❖ Formulae in physics.

❖ Symbols for elements in chemistry.

❖ Irregular verbs in French.

❖ Quotations in literature.

Here is my way of going into the process of memorisation:

1.  Arouse curiosity or anticipation.

2.  Make the child feel the lack of the knowledge.

3.  Present the data in a ready-to-learn form.

4.  Set targets for memorisation.

5.  Help with memorisation techniques and a practice schedule.

6.  Test a lot with quick feedback until the target is achieved.

My area of interest in this book is mostly with the first stage, provoking curiosity, but I think it would be useful to show how it can link into the stages that follow.

So, when I am trying to build that initial interest in my topic, the question I always like to start with is, 'Why do they need to know this?' This is how the internal dialogue continues if I think about times tables:

---

Why do they need to learn times tables? So they can multiply.

Is it possible to multiply without learning tables? Yes, it's just slower.

How would they do that then? By adding a number to itself lots of times.

---

And then it occurs to me that multiplying is, basically, just adding. A special kind of adding, one where you repeat the same action. So, if I learn that $4 \times 3$ is 12, I am learning that $3 + 3 + 3 + 3 = 12$. Then it strikes me as weird that I should find it necessary to learn $3 + 3 + 3 + 3$, but not, for example, $2 + 3 + 4 + 5$, which has a nice regular pattern to it but isn't something anyone ever bothers to learn. Is it just a tradition or a prejudice that makes us concentrate so much on these peculiar repetitive additions? No, we concentrate on them because they come up so much in real life, whatever we are dealing with. If we didn't learn our tables, we would be condemned to performing the same additions over and over.

# Things to Do 8

It is important that children have a real grasp of the concept of multiplication if you want them to attack the task of learning tables with any oomph. And if multiplication is so much a part of life, then it is important to show that fact from the beginning. Here are some everyday ways in which we multiply:

- ❖ Buying multiple packs of something in a shop (e.g. two 2-pint bottles of milk).

- ❖ Calculating our own pay or someone else's (e.g. 35 hours @ £11.25/hour).

- ❖ Estimating our weekly cigarette or alcohol consumption (e.g. 7 x 10 fags a day).

- ❖ Working out dinner money for a week (e.g. 5 x £1.80).

Put children in a position where they need to multiply and get them to feel that lack of knowledge. This is the perfect place for them to be introduced to the new idea – that by learning a few number facts we can answer some of these questions. If you like you can baffle them with your own brilliance, like this:

**Do** Take seven boxes of twelve pens (or some similar sets of items) and show them to the class (preferably use sealed packaging to enhance the idea of a magic trick).

**Say** Who can work out how many pens there are altogether? You can open the boxes if you want!

**Do** Let them see you write down the answer on a scrap of paper, but don't let them see what you have written. When they have finished counting all the pens, produce your answer – proving that you worked it out much faster than them. Repeat the trick by refilling the boxes with the correct number of pens and letting them choose the number of boxes. No matter whether they choose four, five, six or however many boxes, you will always know the total number of pens.

**Say** I can do this because I remember a lot of information in my head. Would you like to see what I know?

**Do** Display the data – show all the multiplication tables, from 1–10 or 1–12, on one screen or poster. Again, you can show off and get them to test you while your back is turned to the board. Once you have got them all fired up, tell them:

**Say** You are going to be able to do this. Just like me. By the end of today, you will learn one line, and by the end of term you will know it all. In fact, you know some of it already.

Research from the marketing industry has shown that a 'you've already done some' approach works wonders.[1] Consider this story. A cafe introduces a loyalty card. The manager is debating whether to give customers a loyalty card where they need eight stamps to get a free coffee, or ten stamps but they get the first two automatically before buying anything. It shouldn't matter because, as you may have spotted, in either case the customer needs to acquire the same number of stamps – eight. But, apparently, you will get a better response from the ten-stamp card with two given for free. Just the fact that we are part of the way to the finishing line is enough. In fact, there is good evidence that if you made it a fifteen-stamp card with five for free that is better still, even though the customer now needs ten stamps compared to eight!

Do the same with the times tables. Show them that they have some 'for free'. The 10 and 11 times tables are easy. Any child within a typical range can learn them in minutes – there's barely anything to learn.

Some children may already be able to count in 2s up to 10 – especially if you have had them doing this for weeks in preparation. All they need to do is keep going – 12, 14, 16, 18 is a simple jump from 2, 4, 6, 8.

The 5s are straightforward too. So are the 9s if you learn one of the many tricks. For example, you hold your hands out, palms facing you. If you want to know what 4 x 9 is you put your fourth finger down. To the left of the gap you have three fingers and on the right of it six – so the answer is 36.

At some point teach them to flip, so they know that 5 x 4 and 4 x 5 are the same. Which means that if they already know the 2, 4, 5, 9 and 10 times tables then they know half of all the others because, for example, in the 8 times table, they already know 2 x 8, 4 x 8, 5 x 8 and so on. This would mean that if they left the 7 times table till last, the only new thing to learn would be that 7 x 7 is 49!

---

1  See Chip Heath and Dan Heath, *Switch: How to Change Things When Change is Hard* (London: Random House, 2011).

There are lots of ways of recording progress. I would favour individual score cards, as some of the wall displays with people's names on them have the usual suspects at the top and the bottom. Doesn't that make the leaders smug and the stragglers hopeless? How about this though: take the whole class of thirty and say that they each have ten tables to learn, so the target *for the class as a whole* is to get 300 tables learned. If brainy Brian learns all his after four days (or has already been taught them by his weekend maths tutor) then Brian's job is to teach as many other people as he can, so that he can help get the class to their total of 300.

Get them doing it in breaks and the odds and ends of time in the school day, practising on the walk home from school, before bed, over breakfast – they need constant reminders. Make sure children know how to explain to parents what help they need. For example, 'I'm only learning the 4 times table for now, and I want lots of encouragement. You test me with this card, and I will try to say the answers.'

You may already be using some of these strategies or others that you prefer. My main point is that a desire to be an interesting and creative teacher must not stop anyone from getting down to the nitty-gritty when the time comes. And the reverse is also true: just because we have to deal with the nitty-gritty sooner or later, that is no reason to give up on creativity, independence and imagination. They are needed more than ever.

So, if you want children to memorise something:

- ❖ Make it meaningful/fun.
- ❖ Tell the children that they must memorise it and show them why.
- ❖ Organise it in a memorable way – preferably several ways.
- ❖ Help them practise recalling it.

If you haven't done this, you can't expect children to retrieve the information. All kinds of things will stick in their memories – anything that their brains happen to take to. It just might not be what you were planning.

# Key Word 3: Memorisation

Memorisation is a forgotten art, even though the storage and retrieval of data in the mind is a vital part of daily functioning, let alone academic progress. It is a mistake to think that we can leave it to computers and phones to remember things for us. Did you know that people who suffer from amnesia (memory loss, often due to accidents) can have problems imagining the future? This is because we use our memories to picture the future. We don't want to leave that to the computers.

The worst way to remember something is to read it or hear it repeatedly, without doing anything. It isn't that this doesn't work at all, just that it is boring and slow. So, don't fall into the trap of telling children things, expecting them to remember and blaming them when they don't. Help them. Make the data *manageable* and *meaningful*.

To make data more manageable, split it into chunks. That is what we do with phone or credit card numbers – reciting them in groups of three or four digits. By doing this, the memory of one item in the chunk sparks memories of the other items in the chunk. Chunks should usually be three to five items.

There is also a thing called the 'Velcro theory of memory'. Memory is thought to be like hook-and-loop fasteners in that it gets its strength from a multitude of weak hooks. So, the sticky strips on a pair of trainers are made up of lots of tiny hooks and loops. The hooks on one strip catch in the loops on the other strip. Each hook–loop connection is weak by itself but lots together are very strong. So, try to associate a new idea with as many different existing ones as possible. For example, someone who wanted to learn that half of a rotation is 180° could learn to associate it with: triple 20 on the darts board, two 90s, the voting age x 10 or the fact that 180 is 'half of three-sixty (360)' and also 'three 60s'. These are the ones that readily came into my head, but the point is that the learner needs to select the ones that strike *her*. It is the teacher's job to kick-start the process.

Colour is very powerful too. Imagining each item in a list in a different colour of the rainbow, or some with stripes or spots, can make them stick in the memory better.

Another technique is narrative. If you want to memorise a sequence of items (e.g. monarchs of England, periodic table, multiplication table, planets in the solar system) make up a silly story – the sillier the better – that links them together. For example: the man who lives at no. 7 has 14 children and 21 snakes. One day he goes out to buy 28 teddy bears which cost £35. He waits only 42 seconds to catch a 49 bus and it takes exactly 56 minutes to get to the bus station. He gets off and buys a bar of chocolate for 63p, which he pays for with 70p in cash. Then he goes to his mother's house which is also no. 7. There is a number on the gate and the door – two 7s. His mother is 84 years old.

Not all information is sequential, like times tables. Another common organising principle is hierarchy. The most obvious example of this is the categorisation of organisms (taxonomy) where we divide all living things into plants and animals, then subdivide them, with those sub-groups being divided again and so on, until you have a pyramid with individual subspecies at the lowest level. In maths, you might divide shapes into polygons, regular and irregular polygons and so on.

When memorising a hierarchy, remember to divide into levels and chunks. At each level of the pyramid, divide the items into chunks of three to five. You may find you can retain more information in a hierarchical format, with its many levels, than in a list, especially if you can visualise it as a chart.

Whether or not you subscribe to the belief that people have different learning styles, you would be a fool to ignore the obvious fact that some people respond strongly to sound and music, and all of us to some extent – that is why children sing their ABC. Encourage kids to make up their own jingles and tunes to remember chunks of data. Practise reciting something as a class with a lively rhythm and intonation. Some children will really get into it – and will remember.

It's funny how a lot of the things that help us remember are the same things that we enjoy. Then again, maybe that shouldn't be a surprise.

# The Cocky Chicken

What is maths anyway? Where does it start? At what point can we say that children are starting to learn it, or that we have started to teach it?

The first thing the child learns is how to count – I think we would all agree on that. Although I would say that there are two separate stages to counting. The first is to memorise the number words in order: one, two, three, four and so on. The next stage is to use these words to count objects around them, like grapes, sweets or marbles. However, it is worth observing whether they do anything with this information – or see it as information at all. For example, you can watch a child count nine peas on a plate, but then find that if you ask, 'So how many peas are there?', you don't get an answer.

This is because, so far, the numbers have been learned as merely a sequence of words, like the words to a nursery rhyme. That is a great place to start, but it is not quite maths yet. What is needed to show mathematical thinking is evidence that the child appreciates the significance of numbers – which is to express quantity and, crucially, to make comparisons between different quantities. For instance, we may want to compare two or more quantities with each other:

We picked forty-eight apples from the old tree and twenty-seven from the new tree.

Leyla has six potatoes, which is more than anyone else.

Or we may want to compare one quantity to a standard, average or ideal:

Hussain is 8, which is quite young to take the bus on his own.

Carla scored all three of her free throws, which is pretty impressive.

When children use numbers to express quantities and compare them, they have discovered a purpose to the sequence of number words that they memorised. An example of this might be if a child can say, 'I have more blocks than you. I have five and you only have four.' It is worth noting the little words that signify the purpose of the counting: *more, not as many, the same, only, too many, not enough* and so on. Learning these key words is at least as important as naming operations, with words like *plus* and *minus*.

To get children to start using numbers in this way (or check if they are doing it already) we can encourage them to compare. Assuming that the child can recite the numbers on the number line, we are looking for two steps:

1.  Count the number of items in a set
    (e.g. pencils in a pot).

2.  Say why that number is important
    (e.g. there are enough for everyone).

The next Thing to Do is an example of a problem you could set to check for these two steps. In maths terms, it is about as basic as you can get, and that is the point. In it, children are asked how many more cookies they need in order to get from three cookies to five. They are then asked the same question in various ways to consolidate that

knowledge. The message for the teacher here is that mathematical knowledge is different from a lot else they are learning and it is natural for them to be hesitant about it. It is different because of its certainty and inflexibility. Two add two always equals four, and can't ever not equal four.

Let's say the child has just discovered that if you need five cookies, and you only have three, you need two more. From this fact, you or I would instantly know all the following things, because they logically follow:

❖ *Next time* I have three cookies, I will need two more to make five (just like this time).

❖ If I have any other number than three cookies, two more will *not* make five.

❖ If I need more than five cookies, *two will not be enough* to get me there from three.

❖ If I need less than five cookies, I will need *less than two* to get me there from three.

❖ If I have three *bananas*, I need two more to make five.

❖ If I have three of *anything*, I need two more to make five.

Let's stop and review this. Because we are adults, we find it forehead-smackingly obvious that if you needed two extra cookies to get from three to five last time, you're going to need two cookies this time to perform the same feat. The same with all those logical deductions I've just shown. But – and it is a big 'but' – it is not necessarily obvious to someone coming to this for the first time. They really may not realise that because you added two last time, that is exactly what you need to do this time or in any situation where you already have three and want five.

This is because the child is still learning when these kinds of 'rules' are 100% reliable and when they are not. Sometimes, it turns out, the same causes produce the same effects – and sometimes they don't. For example, the child is taught that if he asks nicely for a drink of water he will always be given one. One evening he sees his mum pour herself a glass of wine – heaven knows she's earned it – and he asks nicely for one of those. He doesn't get it – even though he asked nicely. So, when does asking nicely work and when doesn't it? Well, it works for water but not wine. Unless you are an adult, in which case it works for wine too. But there might be exceptions to that too. The conclusion seems to be that although certain things make certain other things happen in certain circumstances, other times they don't.

This problem was burrowed into by the Scottish philosopher David Hume. Hume was fat, funny and believed in very little indeed. Famously, he expressed doubts about induction, which is our basic way of learning about the world. What is induction? It is what we have just been thinking about. So, if I prick myself with a pin, I find that it hurts. From this I draw the conclusion that if I prick myself with a pin again, it will hurt again. But why do I believe this? Might there not be a different result next time? Just because something has happened before, why will it happen next time? The natural response to this is something like: 'Because that is how the world works – if something keeps happening it doesn't just randomly stop.'

Bertrand Russell was a much later philosopher, and a master of logic, who put this whole problem succinctly. He takes the example of a chicken who sees the farmer come into the farmyard every morning and knows that this will be closely followed by the farmer giving the chickens their feed. The chicken is smugly certain that feed follows farmer just as day follows night (and pain follows pin) … until one day the farmer comes into the farmyard and wrings the chicken's neck. And perhaps we are a bit like the chicken: cocksure that we know how things work until the day they don't.

Before forming his special theory of relativity, Albert Einstein read Hume (among other philosophers) and gained courage to challenge scientific truths that had stood for centuries and ask, 'What if what we thought we knew is *not always completely true*?'

Hume said there was a difference between two kinds of knowledge. The first kind is the knowledge we get from induction, which includes all our practical experience and everything we claim to know through science. This kind suffers from the chicken problem. The second kind of knowledge has no such problem, and that is logic and mathematics. In our logical and mathematical thinking, our conclusions cannot be overturned in the same way. We know, as I said, that 2 + 2 will always equal 4.

John Stuart Mill, who was one of the most humane philosophers, although frequently wrong, wondered if perhaps we learn our mathematical knowledge in the same way we learn about the natural world: by observation. He reasoned that we know 2 + 2 = 4 because we have done countless experiments and found that it does. This is a disagreement with Hume, who thought that our powers of reason alone tell us that 2 + 2 = 4 and we don't have to test anything.

How does this help you teach mathematics? Because it sometimes helps to remember, as a teacher, that after we have demonstrated that 3 + 2 = 5, we shouldn't instantly expect the child to 'get it' and generalise to every case. If we did we would be presuming that we don't learn mathematical facts by *repeated experiment* (which is called 'induction') as Mill thought, but by *single demonstration* (deduction) as Hume thought. I'm not saying for a moment that a good teacher needs to come to any conclusions about these philosophical issues, or that a child is consciously deciding between inductive and deductive reasoning. What I am saying is that the child's initial uncertainty about whether 3 + 2 will always make 5 is a natural stage to go through.

# Things to Do 9

**Do** Take a group of five children and put three cookies in a jar. Have some other cookies set aside, preferably of a different colour.

**Say** I've got three cookies in this jar. I want to give one cookie to everyone on this table. Can I give one to each person or not?

That's right, it is a very simple problem, but notice that we haven't told the child what to do. She has the tools to solve the problem and, assuming she likes cookies, a motive to solve it. It is time to sit back and see what comes. And what comes may be very little, in which case I will have to scaffold the thinking process:

**Say** I've got three cookies. Can I give one to each person, so everybody has one? … Not sure? … Try it! … Oh dear … What's wrong? … Two people haven't got any cookies … So three wasn't enough …

If this was slow going I would need to run it all again once more, but this time with four cookies. I'd keep going until I'd got them into a routine of counting the cookies, counting the people and saying whether there were enough cookies. Only when they were fully confident with all this would I move on.

Once the children are clear about *enough* and *not enough*, you can intro-duce the next stage, an extension to the not-enough-cookies-to-go-round situation:

**Do** Bring out a box with, say, four more cookies, of a different colour if possible.

**Say** I went to the shop and bought more cookies. Here they are. How many from this box do I need to take out so everyone can have one each?

**Do** If the children can't see the answer straight away, let them distrib-ute the cookies and count how many extra ones they need (the separate colour helps us to remember which cookies were the added ones!).

Our aim is for the children to compare the totals (three cookies from the jar, five people round the table) and calculate the difference – two, of course. But once they have done that you need to work on a conclu-sion. At the very least, the child needs to go back and count how many were needed to get to five. But we can't stop there. We have to get the child to step back, think, generalise and learn. You can use a question like this:

**Say** If we go back to three and do it again, how many will we need?

If I have three bananas, and I want to give one to each person, how many more will I need?

These are adapted from the list of logical deductions in the previous section. The benefit of exploring the implications in this way is so that you can see for yourself what the experiment has proved to the child, rather than fall into the trap of assuming that the child has learned that $3 + 2 = 5$, as this is actually a highly abstract concept, some way beyond the simple fact that we made five cookies by adding two to three.

As well as discovering what your child's thinking is, the questions help you to scaffold their thinking and provide the experiences they will need to consolidate their understanding. So, if you were to ask whether two more cookies will make five when you start off with two, and the child isn't sure of the answer, you can just test it straight away.

By the way, I once ran this activity with several groups for the benefit of two reception teachers and afterwards solemnly asked for feedback. They both frowned. 'Well, I was surprised you put all the cookies back in the jar at the end,' said one. 'I thought that was a bit mean,' nodded the other. I think they had a point, and I noted it for the future.

I then wondered aloud why one of the children had suddenly answered ten when it was wildly wrong. 'It was the way you asked it,' they laughed. 'You said, "You have three cookies at the moment and there are five people, so how many cookies do you want?" And all she heard was, "How many cookies do you want?" – so she said "Ten!"'

There are so many ways the teacher can get it wrong with the very little ones …

# Things to Say 6

*'What do we do next?'*

We want children to be strategic in their thinking. In other words, we want them to know where they are in a process. One way to make them aware of this is to ask them what we should do next and why. There was a point when I wanted to ask this question at every single stage in the belief that this would maximise their appreciation of what they were meant to be doing. But it broke up the flow, feeling rather as if we were playing football and every time someone got the ball I was blowing the whistle and saying, 'Are you going to shoot, dribble, or pass? Why?'

So I learned that you have to pick your moments. The sports tactics analogy is useful. As a coach you want people to get caught up in the game and enjoy it, but over time you want them to grow in understanding of what is happening – to take a mental step back. Sometimes that means stopping the game to make a point, because the example is fresh in everyone's mind. On the other hand, sometimes it is better to go back to a point later, remind ourselves of what happened and then think about why.

The reason I like the wording of this question (and other variations like 'What did we do?', 'What should we have done?') is that it is clear, direct and children usually want to answer it. It appeals to their desire to just get on with things, while at the same time turning their attention to a decision-making framework. This is done in the hope that the child can carry it with them when I am not there to referee.

# Square Are They?

What is a square? It is a question you probably haven't asked yourself lately, and have had no reason to. But if you did have to come up with a definition, what would it be? Here are some properties of a square:

(a)  Two pairs of parallel lines

(b)  Four sides of equal length

(c)  Four angles of 90°

(d)  Four lines of symmetry

Any square will have all these features, but how many of these features does a shape need to make it a square? Which ones do all *rectangles* have, and which are peculiar to squares? For example, if a shape has (a) and (d), does it have to be a square? What about (a) and (b)?

The philosopher Socrates, who is the first person known to have run enquiries of the kind I use in this book, sometimes asked people 'What is … ?' questions. For example, in a discussion about whether a court verdict was just, he would ask, 'What is justice?' This type of question is so typical of him that it is often called a 'Socratic question'. In the dialogues that Plato wrote, where Socrates is the philosophical hero, the character that he is talking to usually crumbles after a while, unable to provide a definition to answer the question. Somewhat triumphantly (Socrates is rarely the neutral facilitator he pretends to be), Socrates asks how it was possible to say whether something was an example of justice if one couldn't say what justice was.

Being unable to *say* what justice is may seem to imply not *knowing* what justice is. But it actually doesn't at all, and Socrates was (if Plato represents him faithfully on the page) guilty of a trick. After all, if Socrates were to ask you, 'What is a square?', you might struggle but you really do *know* what a square is, even if you can't *say* what it is. How can this be proven?

More than 2,000 years after Socrates, Ludwig Wittgenstein had an alternative view, and a method – you may be relieved to learn – to prove that you do know what a square is. His was not the only one, but he is interesting to compare with Socrates because, like Socrates, Wittgenstein was something of a cult celebrity in his time, seemed to mesmerise younger men and combined a passion for logic with a mystical sense of mission. What is more, *Philosophical Investigations*[1], the book that is closest to being his main work (Wittgenstein didn't publish much, rather like Socrates who is thought to have written nothing at all) begins with a discussion of this very issue.

What was Wittgenstein's alternative? Well, he asks you to imagine giving someone instructions to go to the fruit bowl (one containing

---

1  L. Wittgenstein, *Philosophical Investigations*, 4th edn (Chichester: Wiley-Blackwell, 2009 [1953]).

different kinds of fruit and different colour apples) and bring back three green apples. If the person you instruct comes back with three green apples then you can conclude that he or she knows what 'three', 'green' and 'apples' all mean. If the person can carry out instructions with these words in different combinations and contexts then this will be even stronger evidence. This is obvious, of course. But what it tells us is that you don't have to be able to define something in order to know what it means. This has implications for education too. If Wittgenstein is barking up the right tree, then knowledge is related to action – to what you can *do* – above all else. It is not – or, at least, not to the same extent – a matter of what you can *say*.

This activity places action above speech as a way of demonstrating and exploring knowledge, and some children really come into their own because of that. It also provides the perfect opportunity for you to elicit from the participants their beliefs about the properties of a square.

# Things to Do 10

I love doing this one, and it never fails. It is adapted from something I found on the internet some years ago, I think, but I can't seem to find the source again.

**Do**  Take a scrap piece of A4 paper.

Guillotine off five strips from the long side, keep them and throw the rest of the sheet away.

Make the strips about 1 cm in width, as you want them to be thin but not break too easily.

You are now going to walk with these five strips of paper into your class and get them thinking hard for a whole hour.

The first challenge is a bit of a trick. I do it to trick the class into thinking what we are doing is easy. I say:

**Say** I want to cut one of these pieces of paper exactly in half. Who thinks they know a good way?

Of course, they can all do it. Though, interestingly, not all of them volunteer – not sure what that says, but anyway ... The first answer is usually that you carefully fold the paper in half and then cut across the fold. Or some suggest measuring halfway. Let someone execute their plan. Ask them to check that it has been successful (e.g. by laying one piece next to the other). Job done. You now have four long pieces and two of half that length. Now they are ready for the second (real) challenge:

**Say** How many squares can you make by arranging these pieces of paper? (You can't cut or fold them now.)

First ask them to discuss in pairs. At this stage they can see the strips on the floor but can't touch them.

Get each child to say how many squares they think they can make. Common answers are one, two and three. Invite someone who went for a lower number to try first. The child comes to the middle of the circle and arranges the pieces of paper. Other children are allowed to whisper to their neighbours but not interfere with the person working in the centre.

When the child has finished ask him how many squares he has formed, getting him to trace the outlines with his finger for everyone to see. Then ask the rest of the class if they agree about the number of squares. This is an important part of the process, as the group get to grips with the practicalities of the task and the rules. It is likely that at some point someone will say, 'That's not a square,' in response to someone's effort. Use this as an opportunity to pause in the activity.

**Say** Is it a square? Why? Why not?

What makes something a square?

You will hear some truly awful definitions, even if you've just done shapes and have definitions up on the wall. Don't be tempted to step in. Let them reach a consensus. It is a valuable process.

If a child gets stuck, or gives up on their attempt, it is really important that they share their thinking with the rest of the class. That is because the beauty of this task is that so often the final answers are built out of half-finished ones, where one child has an idea but can't make it work, then a second child picks that idea up and does something with it.

**Say** Who can finish Esme's idea?

Who can see what Harry is trying to do? Can you help him?

On many occasions, children will make mistakes about how many squares they have made. A child can step up and help a peer by re-counting correctly, which is good for the collaborative vibe. By watching the children's reactions, you can usually spot the ones who want to adapt or develop someone else's arrangement or to comment on it. Explore one path for a while and then move on to others who want to try a different starting point. It is amazing how little the teacher has to do. Just make sure the children work one at a time and reflect on what is happening.

**Say** How many squares is that? Show us the edges of the squares with your finger!

Would anyone like to try and improve this idea?

Who's got another idea?

Stuck? Tell us what you were thinking!

Here is a typical build-up of answers.

1

Picture 1 is a typical first try and until they have had a chance to play around a bit, plenty of children think this is the best that can be done.

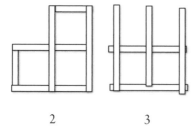

2                          3

In Pictures 2 and 3 the children succeed in making three squares. Note that the sides of the squares are the same length as the shortest strips of paper. Incidentally, when I did this by myself the first time, I thought it was the highest you could go!

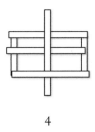

4

In Picture 4, there are only two squares, and it looks as if we need one more strip of paper to finish it off. However, this may spark someone else to think how it can be done. Picture 5 is sometimes thought to be a good one but hopefully someone will notice that these are rectangles. In some ways, these two are actually better efforts than Pictures 2 and 3, even though they form fewer squares, because they are closing in on a better idea – what the children often call the 'window', below.

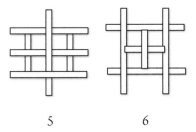

5                    6

These last two are different versions of the same thing. The key leap forward is that the squares are smaller – with sides half the length of the smaller strips. The first time a child put this together, I thought, 'Great, they've found a way to make four!' It was soon pointed out to me that there are five – if you look …

Once the class are satisfied that they have taken the square-building task as far as it can go, elicit from them how they did it. Highlight the fact that sharing a half-finished idea often sparks someone else into completing it, and that everyone improved as they listened to others.

**Do**  Get them to discuss in pairs a new challenge with the same bits of paper.

'How many crosses can you make?' was a good one (the word 'cross' will need to be defined: do the angles between the lines have to be equal?) and so was 'How many triangles?', where again a class of 10-year-olds took the total higher than I thought was possible.

# Things to Say 7

*'What does it need?'*

There are two things a square must have: four straight, equal sides and four right angles. You could say that is eight things, I suppose, but the point is that both of these conditions must be met. A logician would say that they are *necessary conditions*. Neither of them are *sufficient conditions*, because neither is enough on its own to make something a square. Something with four straight, equal sides could be a rhombus, and something with four right angles could be an oblong.

When constructing definitions, or deciding what a word applies to, it is often useful to ask, 'What does it need?' (i.e. the necessary conditions). Once you have agreed on something it needs, you go on to ask, 'Is that enough?' (i.e. is it sufficient?).

In law, this is vital for drafting legislation. Take the example of theft. A necessary condition of theft is taking someone else's property. Another condition is that the thief does not have permission to take it. But that is not sufficient for the charge of theft. Apparently, if the council tows your car away and impounds it, then although they have taken your property without your permission, it is not theft. So, we still need another necessary condition, an exception of some kind, that which allows a private company contracted by the council to take people's property. No doubt this condition exists in the law books somewhere.

Look at the following statements made by children and how a 'needed and enough' analysis can correct them.

❖ 'It's an equilateral triangle. It's symmetrical.'

Do equilateral triangles need to be symmetrical? Is that enough to make it equilateral?

❖ 'It can't be in the 3 times table because it's an even number.'

Do numbers in the 3 times table need to be odd?

❖ 'Round shapes are called circles.'

Do circles need to be round? If something is round, is that enough to call it a circle?

❖ 'It's not a machine because it doesn't use electricity.'

Do machines need to use electricity? Do all machines need it?

❖ 'It's a good story because it has lots of adjectives.'

Do good stories need lots of adjectives? Is that enough to make it good?

On the board, you can list the reasons people give. Wipe off any that turn out not to be necessary, and get the children to say which ones are sufficient.

Sometimes you can have sufficient conditions that are not necessary. For example, a story can be good because it is really scary or really funny (those things are sufficient to make it good), but it could still be good without either of those things (so neither is necessary). Likewise, in law, someone can be a citizen of a country because they were born there or because they have married another citizen. Either of these conditions might be sufficient but neither is necessary.

# Prime Innit, Sir

In doing the research for this book, I have come to the conclusion that there is an easy way to tell the difference between mathematicians and ordinary people: just ask them about prime numbers. Mathematicians love them, for some reason. I have asked them why and they have patiently explained it to me. I can't say I totally get it, but here goes …

A prime number, as you may know, is one that can't be divided by any others – apart from 1 and itself. So 9 is not a prime number because it can be divided by 3, whereas 11 is a prime number because it can't be divided by anything except 1 and 11. (To be precise, by the way, I should say that we are talking about dividing without remainder.)

The lowest primes on the number line are:

2, 3, 5, 7, 11, 13, 17, 19, 23, 29 …

As you can see, there is only one even number in the series – 2. After that, no even number can be prime because we could divide it by 2 – that is what 'even number' means. So, after 2, all prime numbers must be odd.

Why are they called 'prime'? Well, prime means first or most important, and that is a key to why mathematicians like them, it seems. They say that primes are the building blocks for all other numbers because primes are not made out of anything (except 1 and themselves). Other numbers are made out of primes so, sooner or later, they say, it all comes back to primes.

My own favourite numbers are not primes. I like round numbers like 5 and 10, or ones with lots of factors (i.e. numbers you can divide it by) like 12 and 60. A prime like 23 is awkward to use, and especially to divide. But that tells you I'm not a mathematician.

Ever since human beings started exploring numbers they have been fascinated by primes, and interested in finding newer, bigger ones. It is difficult, though, even to this day, to find out if a very long number has any factors. Obviously, we can now get computers to help with the job – which is called factoring – for us. But as the numbers we investigate get bigger and bigger, the number of checks that have to be run get more and more enormous. Funnily enough, the difficulty of this on-going effort to search for new prime numbers is a big part of the attraction.

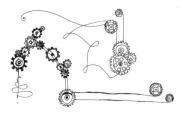

Prime numbers have interesting uses, that's for sure. One of the uses is to stop gear cogs from wearing out. How does that work? Well, cogs will always have tiny flaws on each tooth. These tiny flaws – bumps, scratches, grooves – will wear down the teeth on other cogs when they grind together. So, as the cogs turn, you don't want the same teeth on each wheel to meet each time they go round, because then the little bump on one tooth will press down in the same place on the same tooth on the other cog, over and over again.

For example, let's say you have two connecting cogs, one with six teeth and the other with nine. When the six-tooth cog has gone round three times, the nine-tooth one will have gone round twice – both will be back where they started, and the same teeth will meet up. But if you had one cog of five teeth and one of eleven, the five-tooth cog will have to go round eleven times (and the eleven-toother five times) before the same teeth meet. So the impact of the flaws is spread around.

Evolution 'knows' this too, apparently. Cicadas in North America hatch every thirteen years, or every seventeen years, depending on the species. Both of these are prime numbers. Biologists suspect that these random-looking numbers are not random at all. The cicadas, apparently, are avoiding predators.

Because cicadas are underground most of the time and only emerge to hatch, no predator could rely on them as annual food. Any predator wanting to feed on them periodically would have to synchronise its own life cycle to exactly thirteen or seventeen years, which is pretty unlikely – there are simpler ways of eating bugs for dinner. Whereas, if the cicada life cycle was sixteen years, any predators on a two-, four- or eight-year cycle would get to eat bugs regularly. Whether or not this theory explains the whole thing (and you never know with evolutionary theories), it is amazing that a little insect's body can count to seventeen.

Perhaps the really big use of prime numbers is in encryption. When you buy something online with a credit card, the information you send across the web is encrypted in a secret code. It all gets very technical, of course, but prime numbers are at the heart of it. Basically, if you multiply two prime numbers together (e.g. 7 and 11 to make 77) you get a number that only has these two factors and no others. For someone who doesn't know what those two factors are, the only way to find out is to go through every possibility one by one (or use special algorithms called number sieves).

When you start dealing with very big numbers this process of factoring is just too long to be practical for the hacker. Or so we hope. In theory, I can tell anyone I like the big number I've made by multiplying two primes, but they can't work out which two primes I used to get there. This is important because it means that anyone can put information into my code (because they use the big number to do that), but no one else can then decipher it – because they need the two prime factors to do that.

# Things to Do 11

**Do** Write these numbers on the board: 21, 71, 81, 91.

**Say** Which of these numbers is prime?

A lot of teachers do mental arithmetic or multiplication table drills to open or close maths lessons, or to fill snippets of time too small to start anything else. Throwing out numbers to check if they are prime gets the children working in the opposite direction: instead of thinking, 'What does 7 x 3 make?' they are thinking, '*What* times *what* makes 21?' To work out if something is prime, they have to run through the tables they know.

But the children can become more strategic at this game, and the strategies they develop can be useful for other kinds of problem-solving.

The presenting problem is: 'Is this number prime?' The problem behind it – the problem about the problem – is that we have to go through all the times tables we know to work it out. Or do we?

Let's say the number is 21. (Pretend for now that you haven't already realised that 21 can be made with 3 x 7 and so is not prime.) You can tell straight away that certain tables won't be needed:

❖ 10 times table – because all the numbers in it end in 0, whereas 21 doesn't.

❖ 5 times table – because all its numbers end in 5 or 0.

❖ 2 times table – because the 2 times table is all even numbers and 21 is odd.

❖ 4, 6 and 8 times tables – because they only contain even numbers (because if you multiply any number by an even number the answer will be even).

That only leaves the 3, 7 and 9 times tables. But there is still room for more strategy. Some people might run through the 3 times table first because they know it better or find it easier (others might find the 9 times table easier, perhaps). Or, you might think, the 9 times table has fewer numbers in it (only 9 and 18 before you go past 21) so it is quicker to run through.

Start by revising the prime number concept in the way suggested above, and after a few goes elicit a definition of a prime number, just for reference as the lesson progresses.

**Say** What is a prime number?

The answer should basically be: a number that can only be divided by 1 and itself. A more exact definition would make it clear that we are talking about dividing without remainder, but that may not be relevant for your purposes.

Give them practice with a few simpler numbers like 9, 16, 17 and 18. Check that they know what they are meant to be doing, even if they don't all get the answers right.

Then set them this challenge:

**Say** Make a list of all the prime numbers up to 100.

Bear in mind that the value of this activity is in the strategy. Don't let them start until they have agreed – or at least disagreed – on a strategy. Force the children to be explicit about it, because they won't want to – they'll just want to get on with doing it. Then make a note of the chosen strategy and follow it for a while, reviewing it at intervals. Children need to consider these questions about the strategy:

❖ How can we be sure that *all* prime numbers are on the list – none missing?

❖ How can we be sure that *only* prime numbers are on the list – no false ones?

❖ How can we do only the *necessary* work to make the list – no wasted work?

You might not need to write these up or labour the point – just be aware that this is what the class should be thinking about, and praise any efforts to evaluate the strategy in this way. They should know that they can change strategy at some point, if they have a good reason, and that decision may be the best bit of thinking they do today.

**Do** Have a 1–100 number chart ready, like this:

| 1 | 2 | 3 | 4 | 5 | 6 | 7 | 8 | 9 | 10 |
|---|---|---|---|---|---|---|---|---|---|
| 11 | 12 | 13 | 14 | 15 | 16 | 17 | 18 | 19 | 20 |
| 21 | 22 | 23 | 24 | 25 | 26 | 27 | 28 | 29 | 30 |
| 31 | 32 | 33 | 34 | 35 | 36 | 37 | 38 | 39 | 40 |
| 41 | 42 | 43 | 44 | 45 | 46 | 47 | 48 | 49 | 50 |
| 51 | 52 | 53 | 54 | 55 | 56 | 57 | 58 | 59 | 60 |
| 61 | 62 | 63 | 64 | 65 | 66 | 67 | 68 | 69 | 70 |
| 71 | 72 | 73 | 74 | 75 | 76 | 77 | 78 | 79 | 80 |
| 81 | 82 | 83 | 84 | 85 | 86 | 87 | 88 | 89 | 90 |
| 91 | 92 | 93 | 94 | 95 | 96 | 97 | 98 | 99 | 100 |

**Say** How could you use this to help you?

You may want to let the class haggle over a strategy before they see this, and wait for them to decide they need it. Or show it to them and ask how they could use it, in order to get a strategy out of them. One big chart on the board is fun. Little ones to hand out could also work.

See how they get on. Any class with a decent grasp of multiplication can complete this task. They just need to go through their multiplication tables and cross out the numbers that come up. At the end, the ones not crossed out will be prime. The challenge, as I have been emphasising, is in being strategic – taking the smartest route to the answer and cutting corners intelligently.

As I pointed out, children may or may not discover that the smartest thing is to work out – either at the start or as you go – which times tables you don't need.

Most people start by crossing out the even numbers. This is logical as you can eliminate half the suspects in one go. You can do this whole columns at a time. Then you could go on to the 3 times table. Take a look at this annotated chart:

| | | | | | | | | | |
|---|---|---|---|---|---|---|---|---|---|
| 1 | 2 | ~~3~~ | ~~4~~ | 5 | ~~6~~ | 7 | ~~8~~ | 9 | ~~10~~ |
| 11 | ~~12~~ | 13 | ~~14~~ | ~~15~~ | ~~16~~ | 17 | ~~18~~ | 19 | ~~20~~ |
| ~~21~~ | ~~22~~ | 23 | ~~24~~ | 25 | ~~26~~ | ~~27~~ | ~~28~~ | 29 | ~~30~~ |
| 31 | ~~32~~ | ~~33~~ | ~~34~~ | 35 | ~~36~~ | 37 | ~~38~~ | 39 | ~~40~~ |
| 41 | ~~42~~ | 43 | ~~44~~ | 45 | ~~46~~ | 47 | ~~48~~ | 49 | ~~50~~ |
| 51 | ~~52~~ | 53 | ~~54~~ | 55 | ~~56~~ | 57 | ~~58~~ | 59 | ~~60~~ |
| 61 | ~~62~~ | 63 | ~~64~~ | 65 | ~~66~~ | 67 | ~~68~~ | 69 | ~~70~~ |
| 71 | ~~72~~ | 73 | ~~74~~ | 75 | ~~76~~ | 77 | ~~78~~ | 79 | ~~80~~ |
| 81 | ~~82~~ | 83 | ~~84~~ | 85 | ~~86~~ | 87 | ~~88~~ | 89 | ~~90~~ |
| 91 | ~~92~~ | 93 | ~~94~~ | 95 | ~~96~~ | 97 | ~~98~~ | 99 | ~~100~~ |

The even numbers and the 3 times table up to 36 have been crossed out, which is where the children's knowledge runs out – they don't know 13 x 3. At this point, the smart problem-solver will be looking for a *pattern*. It's not that obvious right now, but if you cross out the next few you will start to reveal it: the fourth row is just like the first row, the fifth is like the second and so on. Using patterns is often faster and easier on the brain than calculating the numbers.

Going back to the times tables you need and those you don't, children may spot a pattern in these too. Assuming you start with the lowest, the 2 times table immediately strikes out half the numbers on the chart. You then do the 3 times table. You don't need the 4 times table because all its members are in the 2 times table and will have been crossed out already. Then the 5 times table needs doing. The 6 times table is even, so doesn't. But 7 does. See whether you spot the pattern if the ones that do need doing are in bold:

**2**, **3**, 4, **5**, 6, **7**, 8, 9, 10, **11** …

They are the prime numbers. In other words, to find prime numbers you only need your prime number times tables. And you can reduce the workload further because inside each of the *prime number times tables* you can skip every other number in the series because it is multiplied by an even number – so it will produce an even number – and will already have been crossed out when you did the 2 times table first up.

So, in the 5 times table you only need to bother with 1 x 5, 3 x 5, 5 x 5, etc. and you can skip 2 x 5, 4 x 5, 6 x 5, etc. In fact, you might have noticed now that there is no point checking 3 x 5 as that will have been crossed out in the 3 times table. And 7 x 5 will have been done in the 7 times table.

The same thing is happening again. It is the prime numbers that are left. So you actually only need to look at the ones where you are multiplying a prime by a prime. And there are not many of those.

For many groups, identifying the prime numbers at all is achievement enough, and to hone their strategy as I have described is beyond them. But keep pulling them back to this idea of strategy and see how far they get. Don't, whatever you do, make children feel that by using a slow and simple strategy they have achieved less. In maths, there are usually infinite ways of getting the right answers, and the one we adopt may be the quickest or it may be the one we are most comfortable with. Of course, we will want to attract them towards a more efficient strategy by demonstrating it to them and helping them master it. But making them feel bad about something that works is not the way to do it.

# Key Word 4: Strategy

This is a way of talking that you might need to build up to, as the children might find the word 'strategy' off-putting if they haven't heard it used before. But it won't take long, as children are good at adopting new words like this. (The day before writing this, a Year 2 class told me that adding is 'commutative' and that subtracting isn't. Luckily, I had learned the meaning of the word about three weeks earlier!)

I talk about strategy and systematic thinking a lot in this book, and here is a tip about how to teach it. At the end of a task, if you ask children which strategies they used and how they got to the answer, they will only mention things that led to the answer directly. So, if they got to the answer by elimination, by logically working through all the alternative answers and eliminating them until they found the right one, then that won't be mentioned by them as part of the solution. Only the last shot, the one that worked, will be seen as having value.

This is no good, because children need to understand that the answer is not everything. Knowing how to get to it, and knowing that you are doing it the best way, is of more far-reaching value. So, the teacher needs to interrupt the task to focus on the question of the strategy: how are we trying to do it? Will that work? Is there another way?

This is why we work as one group, so that the teacher can commentate and draw attention to the process by which the answer is being found. At the end, she can remind the class of key moments that helped them get to the answer (what made a difference, why and when it happened).

If you have multiple groups, the teacher doesn't see enough of the process to make the children aware of the key moments in it.

# Arch Angles[1]

It is 4000 BC. You want to be the envy of all your friends, so you build a great big hut with strong walls. But then you realise you have messed up. There is no doorway. You can't get into the thing. The neighbours think it is hilarious. Angrily, you bash a hole in the wall to gain entry to your new home. The roof collapses and you are now left with a pile of rubble.

Human beings learn the hard way, but they do learn. Next time you will build a door into the hut from the very beginning. You will work out that you need something strong to hold up the roof over the doorway so you get a massive tree trunk to form a horizontal beam. To hold the massive tree trunk up you ram two other massive tree trunks into the ground on either side as supports. Well done. You've invented the doorframe – three straight lines to form a rectangular hole from the ground up to head height.

Next, you move from wood to stone, and shape three huge oblong boulders (best to get some slaves to do it or it'll take forever). That is how Stonehenge was built: a huge slab of rock across the top of the doorway, supported by two other huge slabs of rock on either side. If

---

1  See also Peter Worley, 'Il Duomo', in *Once Upon An If* (London: Bloomsbury, 2011), pp. 151–155.

you build lots of these in a circle you get Stonehenge. This way of supporting weight is called the post-and-lintel system, the lintel being that all-important horizontal bit. In engineering terms, it's not too sophisticated but it works fine.

Scroll forward a few thousand years till you get to Ancient Greece in about 500 BC and the Greeks are building fantastically elegant temples all around the Mediterranean area. They have refined the art of architecture into standards of beauty that survive to this day. However, their engineering concepts have not moved forward one jot from the Stone Age. The incomparable Parthenon, which stands like a sentinel over Athens, is built on the post-and-lintel system: one thumping great slab held up by two posts. All they do is put a lot of these in a row and get a roof on. For all the great leaps that the Greeks made, they contributed precious little to structural engineering.

That had to wait for the Romans. The Romans came up with many amazing things and one of them (we think) was to invent the arch. Instead of laying a long rock over the doorway, they constructed a half-circle out of smaller stones. We are so used to seeing them now that we forget how clever they are. One of the tricks is to use a keystone, a larger central stone, to lock the arch together and stop it collapsing in on itself. Basically, gravity tries to pull the keystone down but it can't fit through the gap between the stones on either side, so it pushes down on them, and they push down on their neighbours and so on, until the last stone in the arch pushes down on the vertical post. The important thing is that arches can carry more weight than lintels because lintels snap if you overload them. With an arch, the weight is carried down into the supporting posts and walls – not borne by the lintel.[2]

The rest of a Roman building is also constructed using arches. Imagine you have built two arches one behind the other to make a thick arch, and then a third and a fourth on the back of those, the arch getting deeper and longer, and before long you have made something that looks like a railway tunnel. This is called a vault. It's a bit dark, so in the side walls you put more arches to make windows. Now it is basically a Roman basilica – forerunner of the Christian church. Many old churches and cathedrals are built on the foundations of ruined Roman basilicas. Basically, medieval architects just put the arches back up.

A Roman arch is semi-circular. It is better than the post-and-lintel system but it still has limitations. So, when the Romans wanted to build

---

2  Solid objects are stronger under compression (squashed) than tension (bent apart). For example, if you snap a stick, the side that breaks first is the one being stretched, not the one being crushed.

high, they had to use masses and masses of stone – look at the colos-
seums: there is more stone in the structure than there is space.

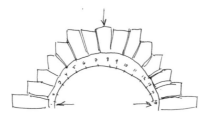

Towards the end of the Roman era they seem to have developed a
stronger kind of arch. Instead of making it round they made it pointy.
This big step would have allowed them to create much higher buildings without using massive great walls, but they didn't get the chance to
develop it far because their civilisation collapsed and they had other
things to think about.

The Arabs, following on, seem to have known about the pointed arch
before the Europeans and there is a bit of a debate about whether the
Europeans learned it from the Arabs or not – not that it matters here.
Why was the pointed arch stronger? Because it was better at conveying
the weight into the posts, because its shape closely followed the direction in which the weight was pushing down.

A pointed arch is made by intersecting segments of two circles over the
middle of the doorway. Huh? That sounds complicated but is simple
when you see it:

This is the shape that visually defines European Gothic architecture. As the centuries inched by, cumulative advances were made by medieval engineers. Although they started by making very narrow arches to hold up huge walls, they gradually learned how to make arches higher and wider. It is relatively easy to tell the age of a cathedral in England or France simply by looking at the width of the arches, especially the ones that span windows.

The medieval Europeans found that, despite their advances, once you had built your arch to a certain height, it would still eventually collapse – the posts on either side would topple outwards under the weight. Their solution was to build an extra bit of wall outside the building to hold up the arch that is supporting the roof. This is called a buttress. The buttresses on later Gothic cathedrals were nick-named 'flying buttresses' because they were beautifully carved and extended out like the folded wings of a resting swan or butterfly.

In the next few centuries, other types of arch became popular, many of them becoming the signature of a certain architectural style or period. One thing they all have in common, though, is that they are made out of circles. Not whole circles but segments of circles. Architectural historians talk about 'centres', so a two-centred arch is made out of two circles.

To draw an arch, you need to work out where your centres are – they are where you put your compass point. To make an early Gothic arch you place the compass point *outside* the posts but level with their height. Mid-period Gothic arches can be drawn by placing the compass point at the top of the post, on the inside edge, to create an equilateral arch. Place the point inside the posts and you form a late

Gothic arch. The most refined is the four-centred arch, with a sharper curve (a segment of a smaller circle) rising from the post into a gentler curve (a segment of a larger circle) to form a continuous elegant line – like King's College Chapel, in Cambridge or Gloucester Cathedral.

In London's Victoria and Albert Museum, there are architectural drawings from two or three hundred years ago, and it is amazing to think that a cathedral like St Paul's emerged onto paper from just a ruler, compass and pencil before it came to life in stone. Its architect, Sir Christopher Wren, despised Gothic architecture and there are no pointed arches in his work. He was among those who popularised the term 'Gothic'. It was initially used as a term of abuse. It had the same meaning as the word 'vandal', both words being the names of Germanic tribes that overran the Roman empire and ground the glories of ancient civilisation into the dust.

But it wasn't all about the West. Mosques from Istanbul to Delhi have beautiful arches and domes, constructed along similar lines to the European cathedrals at much the same time – and, possibly, even by some of the same people! One of the most recognisable symbols of the East is the ogee arch. The ogee uses circles with centres *both inside and outside* the arch so that the line swishes back on itself in an s-shape. Again, easy to draw if you know where to put the compass.

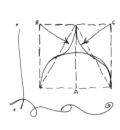 

There are arches with this shape, but we most often see the ogee as part of the 'onion' domes of Russian and Greek Orthodox churches. A dome is actually lots of arches with their points at the centre of a circle. By the way, the word 'ogee' is quite useful in Scrabble. Did I mention that I was better at English than maths?

Since the invention of steel-and-concrete architecture in the late nineteenth century, the distribution of weight is less of a problem because steel encased in concrete is incredibly strong. It has become possible to create gravity-mocking towers, arches and overhangs without having to worry about circles and keystones. It has made modern cities possible but taken some of the art out of architecture. However, along the way we have settled on the ideal shape for an arch, in engineering terms: the parabola. It is ideal because it follows the exact line of 'thrust' – downward pressure – as it descends. Unlike all the traditional arches, the parabola is not formed from segments of a circle, but it has a kind of beauty and is a hugely important shape in mathematics as well as engineering.

# Things to Do 12

We want to blur the lines between geometry and art here. There is a lot of pleasure to be had in drawing some of these shapes accurately and neatly, as you need to activate both the aesthetic part of the brain and the analytical part if you want to get it right. For most children, this will be unfamiliar – they won't associate maths lessons with beauty and proportion, or art lessons with measurement and calculation, so we are expanding both spheres.

Everyone will need a compass. These are not as common as they used to be. The geometry set as a birthday present has definitely been in decline, and the traditional multiple uses of the compass in a secondary school – weapon or tattooing implement as well as drafting tool – have made them unpopular for health and safety reasons. Not to mention the fact that computers draw pretty good circles for us. So, the first thing to do is check that there are enough compasses for the children to have one each. It is possible to create most of the curves by drawing around protractors, because you can see the centre of the curve through the plastic, but you can't do ones of different sizes.

Next, I would recommend squared paper. This will allow the children to concentrate on the elements we want them to, and not use up the whole time trying to draw a horizontal straight line.

It may help to narrate a bit of the history from the post-and-lintel to the arch. One reason to come at it from that direction is that it brings us to the starting point for this task: two posts and a dotted horizontal line marking their height. Anyway, you will certainly need pictures. If you search for any of the architectural terms in this chapter (keystone, ogee, Gothic arch, post and lintel, asymmetrical facade, etc.) you will get lots of pictures to choose from. Try the Robinson Library website to see a clear comparison.[3]

Here are some different ways to create the tasks:

**Say** Here are some arch shapes (you could use photos or diagrams). Can you create them on your paper?

*Or*

**Say** Here are some instructions for drawing arches. Can you match the instructions to the pictures? Can you draw the arches by following the instructions?

*Or*

**Say** You have a sheet with some arches shown on it. Your sheet is different from the other groups' sheets. Write instructions for how to draw one of your arches and give the instructions to another group, who will then use the instructions to draw the arch you have in mind.

Theoretically, it would be a good idea to do one on the board first, although whether you could get hold of a compass that can hold a board marker, I don't know. Perhaps with a normal compass and a heavy pencil on a flip chart you would be OK. Probably best not to attempt drawing it on an interactive whiteboard. If you're anything like me, you'll end up with a scrawl that looks like the signature of an illiterate seventeenth-century peasant signing a confession after torture.

If you wanted to take this more into art territory, you could get children to design the front (or 'facade') of a building by choosing shapes for the windows and doors. Show them plenty of pictures first to fire them up. Then give them an outline on paper as a starting point. This task brings the added difficulty of arranging the windows at equal intervals along

---

3 See <http://www.robinsonlibrary.com/technology/building/details/arches.htm>.

the wall and getting the door right in the middle. That is, if they want it to be symmetrical – most facades are, and before the nineteenth century it was really only incomplete or botched buildings that were asymmetrical. After that they became more fashionable. Still, it is an interesting choice for the designer.

This session also makes a nice lead-in to trips to historical buildings or to the study of certain historical periods (especially the Greeks and Tudors). Once you have been shown how to identify architectural styles, it is relatively easy to do. For example, children aged 3 or 4 can identify something they have never seen before as a church or a mosque as long as it is built in a familiar style.

As this is an engineering concept, there is obviously a science element to be explored. So, in theory, it would be great to test the stability of the different arches by building some and seeing which ones produce the most stable structures. However, to build an arch, you would have to shape each block precisely, with the outer edge being longer than the inner edge. Or you could look at how objects behave under tension and compression – and some of the class may enjoy breaking things in the interest of learning.

# Less to Square

This is the problem-solver to end all problem-solvers. It takes a group of adults about twenty minutes, on average, unless there is a real problem-solving wizard in the mix. A class of children, whatever their maths knowledge, will take longer, and I haven't yet had a class who didn't need prompting at some stage.

The problem isn't the maths. The maths required isn't too taxing for children over about 8, as it is just multiplication, while a familiarity with square numbers does help (so I recommend using it as a revision of that concept). But it does take a lot of organisation, discipline and plenty of application from the class.

A big lesson we can take away from this activity is the importance of recording. It is one of the main gateways to systematic thinking – especially for those of us who are not especially gifted at calculation. Recording results helps us to investigate a problem methodically, doing what we need to do in the most efficient order. This is where the discipline comes in. Working through something systematically until you home in on the answer takes patience.

And then there is the question of square numbers. It is not so easy to demonstrate to younger children why they are important. But they occur all over maths and all over nature. From Ian Stewart's *17 Equations That Changed the World*,[1] here are five of the most famous, and least difficult to understand:

| | |
|---|---|
| Pythagorean theorem | $a^2 + b^2 = c^2$ |
| Newton's law of gravitation | $F = G\dfrac{m_1 m_2}{r^2}$ |
| Standard normal distribution | $f(x, \mu, \sigma) = \dfrac{1}{\sigma\sqrt{2\pi}}e^{-\frac{(x-\mu)^2}{2\sigma^2}}$ |
| Wave equation | $\dfrac{\partial^2 u}{\partial t^2} = c^2 \nabla^2 u$ |
| Einstein's theory of relativity | $E = mc^2$ |

These are all massively important steps in the history of mathematics and science, all still used today, and the basis of all kinds of modern technology. Our sat navs, for example, use Pythagoras' and Einstein's, among many other equations, to determine our location accurately.

And what do they all share? You have to square one or more of the values. So, with Pythagoras, if you measure the three sides of a right-angled triangle, the relation between them isn't clear until you square all those three measurements – and find that the square of the longest side is the same as the squares of the other two added together. And, with Einstein's theory, you have to square the speed of light (that is what 'c' represents, and it is a very long number), then multiply that by the mass of an object, which gives you the amount of energy contained in that object – which is a huge amount. This is why if you release the energy contained in, say, a nuclear explosion, it is a big deal.

The fact is that children need to know this for later. In the short term it has less practical use, perhaps until they get to study area. Try to get the

---

1 Ian Stewart, *17 Equations That Changed the World* (London: Profile Books, 2012).

children to understand square numbers numerically (a number multiplied by itself) and geometrically, and feel confident in playing with the concept, so when they see that little 2 written above and to the right of a number, they will never be intimidated.

# Things to Do 13[2]

You will need a chessboard (draughts boards are the same) but none of the pieces. It can help to have something to write or draw on, so I print out lots of mini-chessboards, four or six to a page, for everyone to work on.

Now, this is one of those sessions that works best if you, the adult, try to solve the problem yourself, and don't read straight on to the solution. Of course, I never manage to hold myself back for long when a book says that, but I hope you can.

Here is the board. The question is:

**Say** How many squares are there?

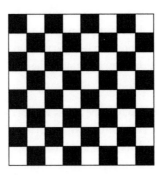

---

2  This activity is adapted from Jo Boaler, *The Elephant in the Classroom: Helping Children Learn and Love Maths* (London: Souvenir Press, 2010).

The first answer you will get, almost certainly, is sixty-four. Ask the speaker exactly how they did it ('Eight times eight', they may say) and why they thought that would give the answer ('Because there are eight rows of eight', for example). Then find out if anyone has a different answer. If not, ask:

**Say** In pairs, think of another answer to the question, how many squares are there?

Sooner or later someone will probably say, 'One. The big square all the way round.' (This is especially likely if they have recently done 'Square Are They?'). Keep digging for more answers and people will start to realise that there are all kinds of squares of different sizes: the big square is made up of eight rows of eight smaller squares, and there are also squares made of seven rows of seven, six of six and so on, all the way down to squares of 1 x 1, of which there are sixty-four, which is where we started.

The task is to count *all* the squares of all types.

**Say** How many squares are there in total, including all these different types of squares?

This is one activity that works just as well even if you let them split into groups to work on it. However, you need to gather and discuss strategy every little while. Find out who is doing what and why. See if the class can divide tasks, with one group counting the 2 x 2 squares, another the 3 x 3 squares and so on.

One question that usually comes up is whether overlapping squares should be counted. For example, we can make a 7 x 7 square starting snugly in the bottom-left corner, and leaving a spare row to the top and the right. Should we say that you can make only one square of this size because those forty-nine little squares are 'used'? I always say no, you count overlapping ones. So that means there are four 7 x 7 squares to discover – one starting from each corner.

There are two other tricky areas that we need to be prepared for, in particular. The first is how we physically count. Doing the 7 x 7s isn't too hard, because we can actually visualise a square slightly smaller than the whole board and shift it up or to the side in our mind's eye (or at least I can, and I always take myself to be roughly normal!). But it gets harder when you try to do the 6 x 6s – you keep losing your place and losing track of what you have and haven't counted. So, ways of marking the board to show which squares you have done can help.

There are various helpful methods. One is to draw the outline of each square as you go. This works but can be messy, which leads to mistakes. A clever alternative is to mark, with a dot, the centre of each square that you have counted. Once you have gone across the whole board, you will need to either erase the dots or start on a fresh board, if you want to count a different size of square.

Some people cut out a square of the size they are counting (e.g. 3 x 3) and lay it across the board, moving it along as they count.

Numerically minded people sometimes do it by calculation. They can see that along the bottom of the board there are seven 2 x 2s overlapping each other. Therefore, if they move up by one small square and start another row, the same thing will happen when they count across – so that is another seven. And there will be seven rows if we go up one each time. So, that makes seven rows of seven 2 x 2 squares, which is forty-nine.

The main point here is that we have to be methodical. We have to know that we are not missing any squares out or counting them twice. So the value of noting and recording becomes really clear.

The second potential problem with this activity is that, the chances are, no one will try to record the results in a systematic way. But if they did, it would look like this, based on what we have so far:

| | |
|---|---|
| 1 by 1 square | 64 |
| 2 by 2 square | 49 |
| 3 by 3 square | ? |
| 4 by 4 square | ? |
| 5 by 5 square | ? |
| 6 by 6 square | ? |
| 7 by 7 square | 4 |
| 8 by 8 square | 1 |

Can you see a pattern? There is just enough information there for some people to see what the series of numbers is going to be. Try looking at it this way:

1   4   ?   ?   ?   ?   49   64

What do 49 and 64 have in common? How do those numbers relate to the single-digit numbers that describe the squares?

What I find interesting is that when presented with the information in this form, several people in a room will quickly jump to an answer, but not before. Others might need a bit more data for the pattern to become obvious, partly because they are not so familiar with this well-known series. Hopefully, when I fill in the rest you will finally see – I have also used the multiplication sign instead of the word 'by' which is an extra clue:

| | |
|---|---|
| 1 x 1 square | 64 |
| 2 x 2 square | 49 |
| 3 x 3 square | 36 |
| 4 x 4 square | 25 |
| 5 x 5 square | 16 |
| 6 x 6 square | 9 |
| 7 x 7 square | 4 |
| 8 x 8 square | 1 |

They are the square numbers and the right-hand column is 'upside down' – 8 x 8 is 64 and so on. But my main point here is that you *don't see the pattern* if you don't record the answers in this order – and children rarely do. The lesson to be learned is that recording data in an orderly way as you gather it can help you glimpse the big picture early on, which means you can then change the way you gather your data – or just stop because you can see what the answer is going to be.

Funnily enough, once they have all these numbers, children frequently add them up wrongly. You probably have your own strategies for mental or quick arithmetic, such as adding the 4 to the 16 and adding the 1 to the 9 or 49, but anyway it is a chance to practise that too!

You should end up with 204, by the way.

# Key Word 5: Resilience

It can be painful to watch an enthusiastic young person set off on a path when you know that it leads to a dead end. The temptation is to step in and steer them back to a trail that will take them on to the answer. I do this regularly. I do it to prevent children becoming discouraged, or to make sure we finish before the bell rings, or because it will take so long for the child to exhaust this fruitless line of enquiry that it is not time well spent. But …

It is vital that children get used to hitting dead ends sometimes. Anyone who gives up when an idea doesn't work is not destined for great things. Electricians, engineers, managers, artists – all of us have to test our ideas. And testing means learning from failure. When we allow children to work through their idea in the same way, we create an opportunity to teach them three things:

1.  To have the gumption to go back and start again.

2.  To eliminate this line of enquiry, which takes them one step closer to the ultimate answer.

3.  To look out for clues as to why they took a wrong step. Could they have known before they started? Or is it just trial and error?

Educators sometimes refer to these qualities as resilience.

Of course, simply getting something wrong will not lead to any epiphanies by itself. We need to be there to praise them for what they have done and show them how it is a vital part of solving the mystery.

# This Land is My Land

Every branch of maths got started in the first place because people needed it for some reason. This is especially true of geometry. The word comes from Greek and literally means 'earth-measuring'. Although it did grow into the noble art of measuring the whole Earth, it started with the dirty business of measuring the earth under our feet: the Greeks needed to measure land.

The idea of selling land was a relatively new one. Land was owned by ancestral right and a religious ceremony was needed to change the boundary of a family plot. But, gradually, the buying and selling of land became more common. And, to this day, you need some basic geometry if you don't want to be caught out. This is because even if you have very regular shaped pieces of land, it is not possible to tell which of two plots is the biggest just by looking at the edges. For example, let's say you have two rectangular pieces of land. One is a long, narrow strip, 15 m x 2 m. The other is closer to a square: 7 m x 10 m. Here are scale diagrams below. Which is bigger?

Well, if you measure all the way round the edge (to find the perimeter) you will find that the long one is 34 m (because it has two sides of 15 and two sides of 2). And what do you know, the second one is also 34 m round because it has two 7 m sides and two 10 m sides. So they are the same size, of course.

Except they are not. Draw them in centimetres on squared paper and you will quickly see they are not. Count and you will see that the long narrow one is made of thirty squares. The broader one is made of seventy squares, so it is more than twice as big. This is worth knowing. If those shapes represent land for farming, you would have more than twice as much room to grow food in the broader rectangle, so it is twice as valuable.

Now, if you had drawn the shapes on plain paper, you wouldn't be able to count squares, and neither can you do that with a patch of earth. But what you can do is to get the same result using the edges. All you do is multiply the length of one side by its neighbour.

We are discovering how to measure, not lines, but *area*.

In the modern world, there are reasons to know the area of land and buildings. If I am opening a restaurant, for example, I need to know how many tables I can fit into the dining area, and so how many people I can feed at the same time, and thus how much cash the business will generate per day. If I want to know how much paint I'll need for the walls of my room, I need to know the area of those walls and check what it says on the paint tin.

However, if truth be told, the practical applications of area are limited for the everyday citizen. Maybe if you rent, buy or sell a lot of buildings you will become familiar with 'square footage' or 'square feet', and quickly be able to visualise the size of a building from a given figure. So, you will know that a large flat or small house might be 1,000 sq ft and a big house might be 2,000 sq ft. But the fact is that you probably don't know the floor area of your home or the room you are in now. And although I said that you need to know the area of the walls you are painting before you choose which paint tin contains enough paint, in all probability you will probably just take a guess. That is one reason why sheds and cellars are full of rusting tins of surplus paint.

Why teach area to primary school kids then? Because it is a fundamental concept in maths that lies in-between lots of others. If you imagine that mathematics is a city, the concept of area is like a major roundabout: people might not visit it for its own sake, but sooner or later they need to go through it to get where they want to go.

One other great thing about geometry (and its close cousin, algebra) is that it is great practice for your brain. You are learning to work out what you don't know from information that you do know. And, importantly, you are learning to distinguish between proving something (in which case, your conclusion cannot possibly be wrong) and simply arguing a point, where you could be wrong but want to persuade somebody that you are right.

If you want children to embrace the concept of area, you have got to go out of your way to make it meaningful. So, either make the effort to devise practical applications or explore it as an abstract concept, or both. But if you just 'cover' it, and move on to the next topic, the chances are you will have one bespectacled boy and his two friends who understand it, and a class full of other children who are all ready to drop the concept into the jumble box in the back of their mind marked 'stuff we did that I don't really get'.

It is a great shame if children don't enjoy geometry. Here are some of my ideas on how to do it. If you come up with better ones, share them with your less experienced colleagues. Or write a book.

# Things to Do 14

**Do** Get the children to draw the two rectangles mentioned above: 15 cm x 2 cm and 7 cm x 10 cm.

**Say** Which is bigger?

Obviously, this question is a trap. The snare is the word 'bigger'. What do we mean by the word 'bigger' here? This perfectly matches the conceptual step the children have to make, because they are having to learn that there are different kinds of bigger: length, perimeter and area. You don't need to tell them that yet. Just show them the shapes and throw them the question. Look out for controversy and let it run its course for a good while. This will engage them thoroughly with the problem, turning a simple, boring question into a genuine puzzle.

Try to gather a range of answers on the board, recording the reasons given for them. Don't allow more knowledgeable ones to cut across and dismiss the more naive ideas – capture them all. And be a bit careful, because there is nothing to say that having a larger area makes something 'bigger' when the perimeters are the same. You could choose perimeter as your criterion of 'bigness'! For example, if you were designing a running track, only the perimeter would be of interest, and the area would be irrelevant. This illustrates the reason for using proper mathematical terminology, which removes a lot of the ambiguity in everyday language.

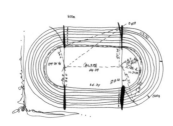

The ideal way to develop the concept is to provide squares to play with …

**Do** Put a set of 12 equal-sized squares, made of paper if necessary, on the floor.

**Say** Arrange these squares to make all the different rectangles you can. Record the perimeter and the area in a table.

One set per class will do it. There are three different shapes of rectangle you can form. If the children know the length of the squares' sides, they should be able to tot up the perimeter. One child can record the results for the class like this:

|  |  |  | Squares | Perimeter |
|---|---|---|---|---|
| 4 | x | 3 | 12 | 14 |
| 6 | x | 2 | 12 | 16 |
| 12 | x | 1 | 12 | 26 |

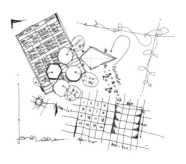

Invite them to come up with a rule that explains this – something like: 'Making thinner shapes gives you more perimeter for the same number of squares. Making fatter shapes gives you more squares and less perimeter.'

This is a good point at which to introduce the term 'area'.

**Say** So far we have been counting using squares. In maths we use centimetres squared and metres squared. Centimetres measure length. Centimetres squared measure area.

**Do** Write the word 'area' on the board, with the symbols $cm^2$ and $m^2$.

**Say** The little ² by the side means 'squared'.

**Do** Give them a dozen pencils or pens of the same length.

**Say** Lay these out to make a rectangle.

How can you use all the pencils and make a rectangle with the biggest area?

What about the smallest area?

This time it is the perimeter that will be constant while the area will change.

Some children may want to experiment with shapes other than rectangles – they should! Praise this impulse and decide whether you can follow it at this point. Make sure that everyone in the group is up to speed with the enquiry so far, before extending it for all of them.

The practical way to find the area of other regular shapes is to divide them into rectangles and triangles, work out the area of each of these smaller, easier shapes and add them together at the end.

If you follow it to a logical conclusion then someone, eventually, will mention the circle. It might not be the same day that you start this activity, or even the same year, but it is where this enquiry naturally leads. So, if you were looking for a way of introducing the area of a circle, and they have already done rectangles and triangles, use this as a starting point to try to arouse curiosity about how far what they know about area can be extended.

Alternatively, you can start the whole thing with football pitches. Ask how big a football pitch is, roughly (a bit of research reveals that the size of a pitch varies). The next question is much the same as that above: which Premiership football pitch is the biggest and which is the smallest? (The dimensions can be found online, e.g. www.openplay.co.uk/

blog/premiership-football-pitch-sizes-2012-2013/.) To answer this question they have to delve into the idea of area. Get them to make scale drawings if they are having trouble.

Another starting point is to ask, 'Which is bigger, this classroom or the one next door?' You could cook up a little story about one class of kids being bigger than another and needing to be assigned to a bigger room. But, anyway, choose any two classrooms that are comparable. In 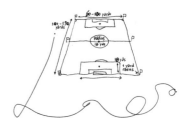 my experience, no two are exactly rectangular – there are alcoves, nooks, cut-aways, pillars and so on. Watch out for them and decide whether to include or ignore them.

Remember, your aim with all of this is to leave the children with a well-rooted sense of what area is and why it might matter. If they get this, they are ready to go on to more complex shapes. If not, don't bother. Go back over this stuff.

# Things to Say 8

*'Which one is the biggest?'*

All the examples above are about which of two things is bigger. None of the questions asks the class to calculate the area of anything for the sake of it. That stage can follow this, as the children calculate the area of various shapes to practise the method. But the introduction to the concept needs to have a sense of purpose, and that is provided simply by the element of competition – even if it is just between two rectangles! For many kids, bigger means better, and they are motivated to find the answer.

Notice that you can split a question like, 'What is the ratio of the smaller square to the bigger?', into two: 'Which square is bigger?' and 'How much bigger?'

This is clear and punchy compared to the longer question. However, exam questions are usually phrased in the longer style, and children will need to get used to interpreting those questions. The interpretation of the text in maths exams is almost a subject in itself and children need lots of practice.

The question of which rectangle is bigger also takes us back to the practical use of mathematics, which is to compare things with each other. We compare them to make them the same or to express the amount by which they differ. Calculating something without comparing it to anything else is to take it out of practical context.

There are other things apart from size that you can focus on. You can ask which is colder, longer, quicker and so on, with much the same benefits.

# Zero the Hero

For a teacher, a story is a great way to get children's attention at the start of a lesson. Simply by saying, 'Once upon a time …' or, 'This is a story about someone called …', you cast a spell across the room. Even boys with a tough image will usually listen – albeit guardedly – to a story to find out how it goes.

Stories don't just grip the minds of children – they lodge in the minds of adults very strongly too. The film, computer game and even sports industries are built on the magic of stories. Before the written word, knowledge and beliefs had to be passed down orally from generation to generation, and they were embedded in narrative. Ancient explanations of the beginning of the universe, the origins of human life and the power of the elements are found in stories – Greek and Norse myths, for example, or the tales of the Old Testament.

Maths and science have their fair share of stories, but this is often forgotten. Many of the leading figures in our intellectual history fought huge personal battles. The lives of Bertrand Russell, Alan Turing, Michael Faraday and Lise Meitner, for example, are compelling but rarely explored compared to those of notable artists and writers. In the next chapter ('Squintasticadillion and One') there is a story about Edward Kasner, who invented a number, and in 'A Bit More Than Three' we hear about Edwin Goodwin who tried to reinvent a number and narrowly failed. To deprive children of these stories is a great shame, because it strips mathematics of its human side, making it seem like a cold, clinical pursuit with no place for passion or inspiration. And that is not how mathematicians experience it. The history of mathematics, while not essential to the understanding of mathematics today, provides an emotional way into the subject.

Some numbers have a story of their own. In *Thinking in Numbers*, Daniel Tammet explains how William Shakespeare, as a grammar-school boy in the 1570s would have been in the first generation of English schoolchildren to learn to use the numerals we use now – before that people used Roman numerals only. The new-fangled 'Arabic' system contained a symbol that did not occur in the old one – 0. This was a number for nothing. And this interest in 'nothing' appears in *King Lear*: 'Nothing will come of nothing' growls the King to his third daughter, Cordelia, who claims that she has nothing to say in praise of her father on the day that he confers her inheritance.

Later in the same play, and throughout Shakespeare's work, there are similar references. The idea clearly fascinated him and he saw the paradox in the concept of nothing that has caused a lot of problems over the centuries. The problem is that as soon as you treat nothing as real you run into problems, because you have made it into a thing, and the whole point of nothing is that it is no thing. Look at this old joke, for example:

---

What is better, complete happiness or a cheese sandwich?

A cheese sandwich. Nothing is better than complete happiness, but a cheese sandwich is better than nothing.

---

This sounds logical but isn't. The flaw in the logic is the double use of the word 'nothing' as if it referred to the same thing twice.

The story below brings together the history of zero and its paradoxical status as both something and nothing, a number and not a number.

Remember that the story you tell doesn't have to have a moral that teaches the curriculum point for that lesson, as long as you can use your story to launch into your topic. The discussion that follows this story leads the children to realise that the number system we use was chosen, not bequeathed to us by a higher power. Other number systems have been, and are, in use. There are reasons why ours, which we inherited from Eastern civilisations, was preferred to the old Roman system. One of them was that calculation was so much easier. Another was its simplicity. This was because it used a place value system, where the value of the symbol depended on which column it stood in. Zero allowed the use of empty columns, which was vital for the system to work.

Thinking about why the system works well and *how* it works helps children to make it work for them.

# Things to Do 15

**Do**  Read this story aloud.

Better still, memorise the key points and alter it to suit your taste or your class. Stop to ask the questions as you go, run a short discussion, and then resume. Write the Roman numerals on the board when you get to them. If you are not confident that they will be of any interest to your class, then you could skip that section.

Here's the story:

Once upon a time, a thousand years ago and more, there was an unhappy number. It was unhappy because nobody used it. People used other numbers all the time, for counting things up and working them out. But no one ever thought about this number. It was completely left out.

One day the number decided that it didn't want to be ignored any longer. So it ran up and down, jumped about, wriggled and called out, 'Hey! Here I am! I'm useful too. I'm a really important number.' But no one seemed to even realise that it was there, and there was nothing it could do.

Time went by, and then the Romans arrived. They brought their armies, and they built bridges and canals and all kinds of clever things. And they brought their own way of writing numbers. They started with:

I, II, III, IIII …

For five they used V. Then they thought of using IV for four, because it is five with one taken off. Numbers taken off went on the left and numbers added on went on the right, so IV, V, VI was four, five, six.

They had X for ten, so nine was X with one taken off: IX. The numbers from nine to fifteen went:

IX, X, XI, XII, XIII, XIV, XV

They used this way of writing to make very big numbers, and since they were a very big civilisation, they needed big numbers to count all the things they had. They could write two-thousand-six-hundred-and-eighty-one like this:

## MMDCLXXXI

Some people found the Roman number system a bit tricky to use, especially when they were doing calculations, but they didn't want to say anything, because you couldn't argue with the Romans.

The unused, unnoticed number thought that maybe the Romans would start using it. 'I can help you,' it cried out. 'Why are you forgetting about me?'

**Say** What was the number and why didn't anyone notice it?

When the number kept on trying to get attention, lots of people dismissed it by saying, 'You're not a number. You're nothing. There can't be a number for nothing. Numbers are for something, not nothing.'

'But what if you count things and there are none?' complained the number.

'We'd just say there's nothing there, of course!' laughed the people, and waved the number away.

**Say** Do we need a number for nothing? Why?

So the number decided to leave home and travel far away, and keep going until it found someone who would use it. It wandered for many years until it came to a very busy, colourful country where everyone seemed to be talking and counting, and buying and selling. It was bewildering and overwhelming. But eventually the number found a group of people sitting under a banyan tree and talking about mathematics. They had an abacus and they used sticks to write their numbers in the dirt on the ground.

'Good morning,' said the number, shyly and politely. 'My name is Zero, and I am a useful number.'

'What are you useful for?' said a lady chewing a snack. She didn't seem all that friendly.

'I am useful for nothing!' cried Zero. Then Zero realised that sounded wrong. 'I mean, when you want to know that there is nothing there, I am the number to use.'

When they heard this a lot of the people started to get interested. They rubbed their chins and scratched their heads. The old men twiddled their beards. The women twiddled their long hair.

Finally, one of them said: 'Welcome Zero. You are exactly what we have been looking for. All of our ways of counting are a bit slow and awkward, but if we use you we can count so much better!'

'Show me!' said Zero, and the little group under the banyan tree showed him how they could count in units, tens, hundreds and thousands by putting a zero in the empty column. Before that they couldn't tell the difference between 21 and 210. Now they realised how to do it.

'So will you use me now?' grinned Zero.

'All the time,' they said, laughing. 'Zero … you're our hero!'

**Say** Is zero a number? How is it different from other numbers?

Is zero a low number? Is it odd or even? Is it prime? Is it square?

# Key Word 6: Objectives

Teachers are often trained to start the planning and delivery of lessons with a learning objective. Observers frequently want to see this objective written on the board and introduced to the class. As you can see, some sessions in this book can lead anywhere. So, what are the objectives?

Well, although objectives can be described in terms of content (e.g. 'Learn how to add two-digit numbers'), there is a risk in only ever thinking that way. There are other kinds of objective, and they are often related to the process not the product. They practise the process of problem-solving, rather than finding the solution to this particular problem.

Here are some examples of objectives to show what I mean:

* Identify a problem that needs to be solved.
* Suggest solutions.
* Evaluate each solution.
* Respond to the ideas of others.
* Give reasons for your opinions.
* Work systematically, not randomly.
* Answer our questions with reasons and tests.

Any of these can be emphasised above the others in a particular lesson: 'Today I want to hear everyone give reasons for their opinions – not just what you think, but *why*?' Children who succeed can be recognised or rewarded.

The actual problem that is being solved in the lesson, and the solution arrived at, are not the main point. It is the process of discovering them that matters. Just as when you go for a run in the park – you go to improve your fitness, not because you need to get to the other side of the park.

And these objectives can't be ticked off after one lesson – any more than going to the park for one run will make you fit (if only!). But regular exercise makes a big difference over time.

Just like physical fitness, the objectives I specify above have a broad application. The skills learned are vital across the curriculum, not just in maths – particularly the habits of working logically and systematically rather than making random stabs at the answer. And not only across the curriculum. Wouldn't they also be important in the workplace?

So, while some of the activities in this book may seem to be nothing more than fun and games, if we wrestle determinedly with the questions that come out of them, we can develop children into independent thinkers. And an independent thinker is a good thing for a child to be.

Of course, for those who see it more narrowly, an independent thinker is a good thing for a child to be in the exam room when no teacher is standing over them to say which method will lead them to the answer.

# Squintasticadillion and One

Children will often say they don't like poetry. They prefer something with a story, something funny, and, anyway, they don't like reading that much. So the poems I write for children usually have a bit of story, a comic tone and are fun to read out loud.

This one came out of conversations I'd had with a few classes about what numbers are, where they come from and where they end:

## Squintasticadillion and One

It's taken quite a lot of work, you know,

But now it's finally done.

I've invented a brand new number:

A squintasticadillion and one.

I thought that a squintasticadillion

By itself was totally new,

But I found it was invented already

By a girl in Timbuktu.

So I scratched my head and wracked my brains

And got out my calculator,

And came up with my unique number

An hour and fourteen minutes later.

I'd like to invent more numbers now

But, well, it's actually pretty tough.

And maybe a squintasticadillion and one

Is enough to count stuff.

---

The question lurking underneath it is: are numbers invented or discovered? Do we create them, as we create new words or new machines? Or are they already there, waiting to be found, like dinosaur fossils or America? For now, the important thing to focus on is how the children try to answer the question of whether you can invent a new number.

One place where children get stuck is in thinking that a new number needs a new name. Each time we move one place to the left, and put a 1 in a new column, we have to decide what to call it. Sometimes we have a completely new word, as when we move from hundreds to thousands. But, for some reason, we don't do that in the column that follows, or the column after, giving us 'ten thousand' for 10,000 and 'hundred thousand' for 100,000. The Indian system does it differently, giving the name *lakh* to 100,000. They would see our million as ten *lakhs*, because they have no separate word for million, and for them ten million is a *crore*. Likewise, many Europeans would call 1,000,000,000 a milliard, whereas those in the US and UK (officially since 1974) call it a billion, a term which was previously used for a million millions over here. The point is that the names of numbers are not universal, or even necessary. Scientists often use $10^2$, $10^3$, $10^4$, $10^5$, $10^6$ and so on for 100, 1,000, 10,000, 100,000 and 1,000,000 to cut out all the confusion and save ink.

Because the place value system is repetitive, we can represent ever higher numbers simply by adding columns and filling them with one of our ten digits, 0 to 9. The meaning of that digit is determined by the place where it stands. In the Roman numeral system, as we saw in the last chapter, this wasn't so. To make higher numbers they needed extra letters to stand for them. That is one more reason why the system was abandoned, gradually, for the super-duper new 'Arabic' system, which – as mentioned in 'The Real Deal' chapter – in Arabic countries was called the Indian system because that is where it actually came from.

Here is a good story about someone who might have invented a number. Edward Kasner (1878–1955) was a mathematician at Columbia University who decided that he needed a new number. Why? Just to interest children. So he thought that a 1 with a hundred zeroes after it should have a name:

---

Words of wisdom are spoken by children at least as often as by scientists. The name 'googol' was invented by a child (Dr. Kasner's nine-year-old nephew) who was asked to think up a name for a very big number, namely, 1 with a hundred zeros after it. He was very certain that this number was not infinite, and therefore equally certain that it had to have a name. At the same time that he suggested 'googol' he gave a name for a still larger number: 'Googolplex.' A googolplex is much larger than a googol, but is still finite, as the inventor of the name was quick to point out. It was first suggested that a googolplex should be 1, followed by writing zeros until you got tired.[1]

The nephew's definition of the googolplex was insufficiently precise for Kasner, so he decided instead that a googolplex should be 10 to the power of a googol – that is, 10 multiplied by itself googol times. As the author points out, it is almost unimaginably long:

---

You will get some idea of the size of this very large but finite number from the fact that there would not be enough room to

---

1  Edward Kasner and James R. Newman, *Mathematics and the Imagination* (New York: Dover Publications, 2013 [1940]), p. 23.

write it, if you went to the farthest star, touring all the nebulae and putting down zeros every inch of the way.

---

Incidentally, Kasner's semi-serious work went on to inspire the name of a multinational company based in the US. Yes, Google. They spelt it wrong, then stuck with it. And then they called their HQ in California The Googleplex. Groovy guys, huh?

So, that is a story about someone sort of inventing a number. For a story about people discovering one, turn to the next chapter.

# Things to Do 16

**Do** Read the poem aloud a few times.

Sometimes I give each child in the class a line to remember, so as I read a line, I point to a member of the class to memorise it, going round the class in order. If possible, I give two people sitting next to each other the same line. When I reach the end of the poem, I tell them to turn to their partner and recite the line as practice. Then the class recites the whole thing back, with each line coming from a different person. They enjoy this challenge and it helps to embed some of the workings of the poem. You can also project the poem onto the board at the end.

**Say** What do you think of the poem?

Do you think the person speaking in the poem invented a new number?

Your class discussion, especially at primary level, is likely to centre on names, as I pointed out in the introduction to this task. One thought experiment to explore this idea goes like this:

**Say** Imagine we have written a number that starts with a 1 and then all these zeroes that you see on the board … and imagine I have written a line of zeroes that goes all along the wall, and out of the door and down the corridor, and round the playground and then all the way back up into the room and onto the board … and these zeroes I'm writing now are the last ones. There. I've finished. That is the whole number. As you can see, it is very long. Imagine someone wanted to work out the next number after this. Could they do it?

Given the chance to discuss among themselves, children don't always find it obvious that all you have to do is erase the last 0 in the number and replace it with a 1. Admittedly, you should find a few kids who realise this, but it may take them a while to persuade the others. If this is eventually agreed by everyone, you could move on to ask them if they have just invented a new number, and draw out the implications: is it easy? Can anyone do it? Is it impossible?

**Do** Tell the story of the googol.

**Say** Did Kasner invent a new number?

What I would hope is that pupils start to see that numbers are under our command – we should not be tyrannised by them. Even if they exist independently of us, and all human thought, it is up to us how we represent them – and the very elegant system that we have inherited from our ancestors can cope with anything. Then again, in the next chapter we'll see that maybe it can't …

# Key Word 7: Auditory

Reading a poem aloud is great for some children who – whatever their reading ability – feel confident and comfortable with messages in auditory form. Perhaps their brains are wired that way, perhaps it is the way significant messages are transmitted in their family or culture, perhaps they are dyslexic, or highly musical, or too lazy to read. But there are usually a few in any group, and you always get a better response from them if they are allowed to just listen – not to a never-ending ramble, but to something clear and punchy.

If you have never come across any theory of learning styles, it is worth taking a look. The idea is that individuals have preferred ways of learning. For example, some may prefer to experience new ideas visually, with pictures and diagrams. Auditory learners, on the other hand, like to hear them and other people prefer hands-on experience.

In actual fact, there isn't much evidence that adapting your teaching style to suit the learning styles of your students will make any great impact on their learning. However, you may feel from your own perspective that there are some ways of teaching that don't help you. A very clear example is with learning a foreign language. Some people can pick up foreign languages by listening to them, rather than actually studying. I can't. Barely a word will stick in my brain that way. I have to see the word written down, and analyse the way it sounds by getting someone to repeat it quite a few times. So, I'm sure there is something in the idea of learning styles.

I simply use these ideas to add variety to my lessons and different ways for children to engage. If I can come up with a visual, auditory and kinaesthetic element to each lesson, then I am doing well. And that is why I put the emphasis on listening to a poem rather than reading it. They read all the time.

# Beauty Secrets

I'm going to introduce you to a number called phi ($\Phi$) – pronounced like the first syllable of 'final'. Here is a glimpse of this special number:

1.6180339887 …

What is so special about it? Well, one of the things to know is that it is irrational. Irrational numbers can never be written down, so the dots at the end of my version of the number above indicate that this is just the beginning of the number. I can't finish it because it has no end.

This is different from a recurring number, by the way. A recurring number is what you get when you divide 10 by 3. As you probably know, the answer is 3.3333 recurring – in other words, the 3s go on for ever. However, there is another way to express the answer: $3^1/_3$ or $^{10}/_3$, which means that it is rational. Irrational numbers, on the other hand, can't even be expressed as fractions, and they never recur, no matter how long you go on. They are that weird.

Weird they may be, but they are common. In fact, they are more common than rational numbers. Infinitely more, because between 0 and 1 there are an infinite number of irrational numbers. However, attention has focused on a few in particular for many centuries, stretching back to ancient times, including the one above. We will see why in a minute.

First, take a look at this sequence – probably one of the most famous sequences in the history of maths. If you haven't seen it before, try to work out what the next number must be:

1, 1, 2, 3, 5, 8, 13, 21, 34 …

It is the Fibonacci sequence and the next number is 55 – because each number is made by adding the two before it together. We add 21 and 34 to get 55. (By the way, you may notice that the first 1 has nothing before it – so you add 1 and nothing to make the next 1!). There is a connection between this string of numbers and phi. Divide any number by the number just before it – for example, 13 by 8. It's 1.625. That is close to phi. And the higher you go along the Fibonacci sequence, the closer you get to phi when you divide two neighbouring numbers.

Here is the first interesting point to ponder about it: although no one has ever been able to write phi down (because it never ends) we can still see it. It looks like this:

Why is this phi? The whole line is phi (i.e. 1.61 … ) times the part on the left, and the part on the left is phi times the length of that on the right. The line is divided at a unique point, the only one where the relationship between the two lines on either side is the same as that between the longer line and the whole. This point of division is called the golden section.

If you take the two parts of the line and form a rectangle (by adding two opposite sides) you have constructed a golden rectangle. This has a very special property. First, use one line to divide that rectangle into a square and a leftover rectangle. The leftover rectangle will be smaller but will have exactly the same proportions as the one you started with.

You can then draw another square inside the little rectangle to perform the same trick again … and again …

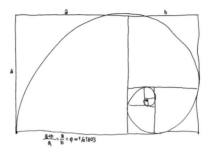

$$\frac{a+b}{a} = \frac{a}{b} = \varphi \approx 1.61803$$

According to tradition, artists have used this method of dividing lines and spaces because it is the most harmonious, balanced and beautiful proportion. In fact, the name phi is said to have been inspired by Phidias, the mastermind of the Parthenon in Athens. Browse the internet and you will find plenty of websites claiming to explain how many of the masterpieces of Western and Eastern art are constructed according to the golden section. However, if instead of taking the author's word for it, you start to divide lengths by widths and so on, you may struggle – as I did – to discover for yourself the golden number phi in any particular building or painting. And if I tell you that the conspiracy classic *The Da Vinci Code* claims that phi is at the heart of the pyramids, the *Mona Lisa* and various other great works, you may get an idea of the danger of being too credulous.

The only writings I have seen on phi in art that attempt a proper mathematical approach are sceptical.[1] However, some of those authors might be being too literal. If the calculations for phi in paintings are only roughly right, does that disprove that the human eye is pleased by it? Wouldn't it be more artistic if these golden proportions floated just beneath the human figures and landscapes that fill our galleries? And perhaps that was what the great artists did – hint at the golden proportion rather than treat it as a formula for beauty.

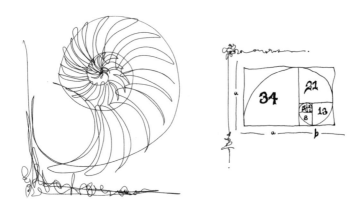

What is not in any doubt is that the golden section, and the Fibonacci series to which it is linked, can be found all through nature.

It can be found in pine cones, petals, tree branches, sunflower heads, snail shells and the human body. And in the reproductive statistics of rabbits. In some cases, phi seems to be an evolutionary solution of some kind. In others it might just be a coincidence. But not only are many life forms carriers of the great number but also spirals found in galaxies and hurricanes seem to be rooted in phi. It is as if it were a

---

1   For a convincing, if slightly heartless, debunking of the issue try George Markowsky, 'Misconceptions about the Golden Ratio', *College Mathematics Journal* 23(1) (1992): 2–19, which can be found at <http://www.math.nus.edu.sg/aslaksen/teaching/maa/markowsky.pdf>

recipe for creating structure and growth. No wonder it strikes our eye as beautiful in some fundamental way.

And one last point: it crops up across the landscape of mathematics too.

This is one of the areas where pure reason and mysticism touch for a moment. For some people, the presence of this number throughout the universe feels like something divine, the fingerprint of God, a proof that creation is not just blind chance. For others, it is proof that no God is necessary to explain the wonders of existence – that the wonders of existence intertwine and interconnect and create themselves over and over again, endlessly. Personally, I rather like the fact that neither religion nor science can explain these patterns completely. Not yet anyway.

# Things to Do 17

**Do**  Start writing up the Fibonacci sequence.

**Say**  What is the next number in the sequence?

Pair the children up to discuss it. If they come up with an answer they should keep it to themselves for now. Go round the groups and if anyone has cracked it – make sure you give them plenty of praise but ask them not to spoil it for others who are still working on it. When the groups report back, get some ideas from those you know are only halfway there, and praise their progress. Then hear from the group that got all the way there to complete the answer.

If no one has got there, that is ideal. This is when you can guide them into good thinking habits. Give each pair or group a different set of three numbers from the line and see if they can find any connection. Sooner or later, each group will spot that if you add the first two you get the third. When they report back, they will hear that everyone (perhaps with some help) has found the same thing.

The children simply need to see that the next number will be the product of the two before it – just as each one has been up until then. Explain that this number sequence is named after an Italian mathematician, Fibonacci, who wrote about it. He was describing rabbits breeding, as it happens. At this point you could go into the Fibonacci series in nature, if it is a topic you would like to touch on, but otherwise you might want to skip straight to art.

**Do** Show Seurat's *Bathers at Asnières*. Pass round calculators.

**Say** There is a direct connection between the numbers and the pictures.

Divide one of the numbers on the list – any you like – by the one on the left.

As they do this they will be homing in gradually on 1.618, which is an approximation of phi.

**Do** 1. Measure the height of the painting.

2. Starting from the left, measure that distance along the bottom of the painting and make a mark.

3. Draw a vertical line up from the point and, of course, you have made a square – with a rectangle left over on the right.

4. Divide the height of the left-hand edge by 1.618 (371 cm on the actual painting) and measure up that distance from the bottom – mark that point (229 cm on the original).

5. Draw a horizontal line across until it meets the vertical.

The two lines should intersect just about at the eyes of the main figure of the boy sitting on the edge of the bank. The artist did this to draw our attention to his gaze and to make him feel 'central' in terms of importance. Putting him in the actual centre would have made the picture flat and bland.

Distribute sheets of paper that have this subdivision drawn on them and encourage the children to make some more, based on the golden section. They can do this by measuring the short side of any of the rectangles they have formed and then bisecting the other side by that distance (forming smaller squares each time). If they draw two or three more lines in this way they will end up with a grid that looks like the illustration at the top of page 159.

Ask the children to compose a picture of their own. They can turn the paper to portrait position. They can draw a landscape, a human figure, anything, but they should position the most important parts (trees, flowers, eyes – whatever they choose) on the 'corners' – the intersections of those lines.

Alternatively, they could colour the lines and squares as Piet Mondrian did. There is a debate about whether he used the golden section or not. No amount of measuring the paintings seems to solve it. It might be fun if one group composed their pictures around the golden section and another didn't, and see if a third party (e.g. the teacher next door) can tell them apart.

I have been wondering whether there is any advantage to knowing any of this, or teaching it. It isn't going to help you measure up some shelves, complete your tax return or check whether your restaurant bill is correct. It is very much in the realms of pure unadulterated maths – something for the last day of term in sixth form to entertain some nerds.

But … but … somehow I think it could be of value.

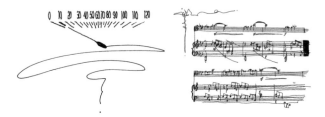

Mathematicians – and I mean the ones who visit the outer edges of human knowledge – tell us of the sheer beauty of numbers, of how the vastness of the number system is not empty and repetitive but teeming with interest and life. And we can't see it. We can't see what they see because we are simply not intelligent enough. Or, if you don't like that idea, it is because our intelligence lies elsewhere. But I believe this just might be a way of giving children a glimpse into that dimension, one of the few areas where its marvels are visible to the 'naked eye' of the ordinary brain.

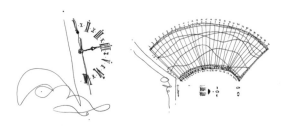

This is how I would try to make a lesson out of it. In practical terms, I suppose this session is most likely to succeed with top sets or other selected groups. But it doesn't have to be only them. You don't need to be good at calculating to spot the relationships between the numbers in the Fibonacci sequence. But you do have to understand the task, adopt a strategy and believe that you are capable of seeing a connection. That is all about attitude. So, as I said, it will be easiest with kids who have the right attitude – but isn't everything?

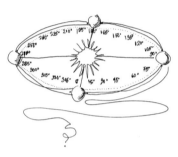

# A Bit More Than Three

Recreating the process of historic discoveries is a wonderful thing: Newton's apple falling from the tree, or Archimedes' bathwater spilling out onto the floor. While these stories are almost certainly myths, they illustrate beautifully the eureka moment. This moment is so exhilarating and gratifying that we should not forget it when trying to teach. Not every lesson of the day can measure up to those heights, obviously, but we have all witnessed the way a 'penny-dropping' moment can transform someone's face into delight.

I think too that it is valuable for children to get a sense of the whole wide landscape of human endeavour: what has come before us, how much has been achieved and how amazing some of our ancestors – with their quill pens, candles and analogue data – really were. We are trying to emulate them in our own way, and to see how all the knowledge we now have has been assembled over centuries – sometimes in flashes of insight, sometimes piece by painstaking piece.

In 'Beauty Secrets' we came across the idea of irrational numbers. These are numbers that can't be expressed in digits. They go on forever, without repeating themselves. Tantalisingly, we can never really grasp them, only approximations of them. In this chapter, we are going to be looking at an even more well-known irrational number: pi ($\pi$).

What is pi? It is the relationship between the diameter of a circle (the distance across) and its circumference (the distance round). That relationship is the same for every single possible circle; it is part of what being a circle is. And the relationship is roughly 3.142. In other words, if you measured across a circle with a piece of tape, that length could be wrapped round the circle slightly more than three times.

As with the magical 'golden' irrational number phi, it is hard to say when human beings first knew about pi, because we don't know how much they knew, and it is hard to say what counts as 'knowing' a number when that number cannot be expressed numerically. We do know that it has been calculated to more and more places over the centuries, and during that time we have become more aware of its properties.

Pi is alluded to, in a way, in the Bible: 'And he made a molten sea of ten cubits from rim to rim, round in compass … and a line of thirty cubits around' (1 Kings 7:23). That would give you a value for pi of 3, and no more. As ever, a holy text is not best used as a scientific textbook. But 3 was used as a rough number by lots of our predecessors. After a while, they realised that they needed to be more precise.

The name pi was first used in 1706. But, before it had a name, generations of mathematicians sought to calculate the ratio between diameter and circumference with ever more accuracy. Here are some previous known 'records' for calculating pi:

| | |
|---|---|
| Babylonians | $3\frac{1}{8}$ on a clay tablet found at Susa |
| Egyptians | 3.16049 (if you translate their description into decimals) |
| Early Greeks | 3.1622 (because they took it to be the square root of 10) |
| Hellenic Greeks | 3.14166 |
| Archimedes | less than $3\frac{1}{7}$ but greater than $3\frac{10}{71}$ |

The Romans used Greek mathematics, on the whole, and after their empire collapsed there was barely any progress in the West. By the third century, a Chinese mathematician had got as far as 3.14159, which is pretty good going. A succession of Indians published values of pi, including 3.1416 in the year 499 and 3.14156 in 1114. Our winner, though, is Tsu Chongzhi, a superstar Chinese astronomer and mathematician, who brought it to 3.1415929 – correct to 6 decimal places and the best effort for the next thousand years.

Interestingly, the values of pi that each of these civilisations reached closely match the values of pi used for practical purposes – by structural engineers or designers, for example – to this day. Maybe that is why they went no further. Another reason was that their methods were very time consuming.

As far back as 1424, the Persian Jamshid Masud al-Kashi realised the futility of extending that line of digits on the right of the decimal point. He calculated pi to 16 decimal places and felt that was far enough. Using that value, he found that if you were measuring something with a circumference 600,000 times that of the Earth, your margin of error would be 'the thickness of a horse's hair', which was a standard old measure in Persia, now thought to have been slightly less than a millimetre.

The fact that it is pointless, in practical terms, to go on and on calculating the value of pi has not deterred everyone in the mathematical community. And once machines had been invented to crunch the numbers, they could be set to work while the humans sat back and waited for those strings of digits to be spat out. The first time a computer attacked the problem, in 1947, it took seventy hours to reach 2,037 decimal places, beating the previous best produced by two mere humans of 808 places. Fast forward to 2013, and someone somewhere claimed a record of 10 trillion digits. I haven't checked their work, so I can't vouch for its validity, but I don't suppose the record will stand for too long.

This potted history, however, has so far omitted the contribution of Dr Edwin J. Goodwin, of Solitude, Indiana. Let's hope he will never be forgotten. In 1897, he announced a 'new mathematical truth' with world-altering consequences. He offered the use of his insight to educational institutions in Indiana free of charge, while the rest of the world would have to pay a royalty to use it. Goodwin persuaded his local representative in the Indiana General Assembly to submit his new truth as a bill.

Unfortunately, his new mathematical truth included the startling proposition that pi should henceforth be 3.2. Regarding pi as 3.2 confers a number of advantages, not least that it allowed Dr Goodwin to solve a puzzle that had frustrated mathematicians for centuries – how to square a circle (in other words, to construct a square with the same area as a given circle, which sounds straightforward but apparently is impossible). The key problem with regarding pi as 3.2 is (as a visitor to the Indiana Assembly on the day the bill was put up for debate urgently pointed out) that ... it is not 3.2. You might as well decide that 12 x 12 is 150 because it makes life easier. Hurriedly, the debate on Dr Goodwin's bill was postponed till a later date. As of 2014, the postponement appears to be indefinite.

# Things to Do 18

The point at which people are introduced to pi is sometimes the point where they disengage from maths. A bit like the point at which I heard Radiohead's fourth album and concluded that I had followed this whole thing as far as I could go, and it was now just alienating and obscure. But people who jump ship when pi gets on board are needlessly panicking. Here are some ideas to steady the ship.

First of all, beware of introducing circumference and area of a circle to anyone who is not confident with the ideas of perimeter and area for other shapes. Your time will be better spent consolidating those concepts than burying them under more half-learned ideas.

But hopefully you will be standing in front of a group of people who are ready for the journey.

**Do** Give the class a set of rectangles and triangles on squared paper, and a table with two columns to record results.

**Say** Find the perimeter and area of all of these shapes and record the results on the table.

**Do** Hand out cylinders of various sizes – like old toilet rolls or other tube packaging. Demonstrate how they can stand the cylinder on its end and draw round the base to make a circle.

**Say** What are the area and the circumference of the circle you have made? How can you find out?

Try to elicit ways of doing it from the class, getting them to fiddle with the problem in pairs.

To get a measurement they will probably have to go back to the cylinder. Assuming they don't have a tape measure, they could *wrap a piece of paper around the cylinder*, mark the point where the paper meets itself, then measure how far that is from the edge by laying the paper flat again and using a ruler. Alternatively, they could make a mark on the side of the cylinder and *roll the cylinder along a ruler*, noting the distance between each time the mark reaches the bottom as they roll.

Then, as best they can, they should *measure the diameter*. To be accurate, you need to measure through the centre of the circle, but this will probably not be marked on the base of the cylinder. You could show – or elicit – that the diameter will always be the longest line from one side of the circle to the other. If you miss the centre slightly, your measurement will come in under length.

You can also construct a square exactly around the circle and the square's centre will also be the circle's. Then measure through this.

**Do**  Get each child to give you their measurement for circumference and diameter, and collate all the data into the same two-column table as before.

**Say**  Which of these columns has numbers that are always bigger? How much bigger?

Children may not know how to answer the question, 'How much bigger?' So you could prompt: 'Is it double?' 'More than double?' 'Ten times as big?' 'Less?'

Let them use calculators to check. They can blunder their way as long as you are there to steer them back on course. Perhaps like this: they take any diameter measurement and multiply it by 2 and then by 10. They then compare the answers to the circumference measurement. Soon they will establish that the figures for circumference are between 2 and 10 times the figures for diameter. Then they can narrow it down, multiplying by other numbers until they find – if the measurements are even roughly right – that … the circumference equals the diameter multiplied by about 3, or a bit more than 3, all the time.

At this point the class is at about the same point in history as the Bible, and catching up with the Sumerians. You can, if you choose, fast forward a few centuries and say that if you do really accurate measurements you find that the circumference is always 3.142 times the diameter. But not exactly 3.142.

If you can, construct a circle whose centre is marked and has a measurable circumference (e.g. a wheel). Try to make the circle big, because then the measurement errors will be a smaller proportion of the overall figure. Compare the circumference and the diameter and you should obtain a result very close to pi.

You can explain that many people have tried to pin this number down but have found that it just goes on and on, and now it is believed to go on forever. When we calculate, we use as many digits as will fit on to the calculator, or fewer if we don't need to be too accurate.

From this point, you could *move on to area*. If you draw circles on squared paper, children can estimate the area by *counting squares*, and parts of squares, and recording their answers in the same way as they did with circumference. However, this time they will need to record the *radius*, which is the distance from the centre to the edge – half the diameter. Follow the same procedure of comparing the two sets of figures. This time the pattern will be harder to discern, so don't expect them to 'discover' it. Let them have a look at it, and then be ready to come in with the formula. This is normally expressed $\pi r^2$. This means that if you multiply the radius by itself (so if your radius is 6 cm you do 6 x 6) and then multiply that answer by pi, you should get the area figure.

It is not essential to work through this process for circumference and area. The main point of the exercise is to introduce the children to pi, and to help them see where it comes from and how it was gradually uncovered over the years. If they find it weird, then at least they can feel that this weirdness is in the number itself, not in a failure to understand on their part.

# Things to Say 9

*'Weird!'*

Back in the Introduction, I said that the teacher should model the right behaviour for learning. One way to do this is to admit that things are confusing, sometimes, but they can still be understood before too long. Failing to acknowledge that a new idea is, well, weird, can make the children feel that the weirdness is in them.

That is why I am urging people to take this long route and not just bung in a weird little number, with its weird little name, without taking into account the effect it will have. That is like introducing a blue Martian with a great big eye as a new member of the class, and expecting no one to make fun of it at playtime.

Teachers may have grown used to teaching pi, and lost any sense of wonder or confusion about it. This is a shame, because to see it as a mere tool for generating answers to maths questions is to leave out the interesting bit of the story – which I have tried to fill in in my introduction to this chapter. And there are a lot of people out there who need to understand concepts fully, and put them in a wider context in order to use them. So, I recommend that pi is introduced in all its unsettling, irrational, inevitable, mind-stopping weirdness. Or at least as much of it as you can muster on the rainy Thursday morning that you happen to be teaching it to young people who fish out their mobile phones every time you turn around to write on the board.

# Inbetweeny Bits

Fractions are the bogeyman of maths. My older sister scared me with them as something I would suffer when I got older, like BCG injections and puberty. Mentioning them in a classroom wins you groans, sighs and slumped shoulders.

Why? Well, the perception is that fractions are 'hard'. Perhaps they are. Look at this fraction calculation:

$$\frac{1}{3} + \frac{1}{4}$$

The first thing that is hard about it is ... that it is *impossible*. You simply can't add a third and a quarter together. That is kind of off-putting. So let's take an easier example:

$$\frac{1}{3} + \frac{1}{3}$$

Phew. Most of us can quickly see that the answer is two-thirds. It is really the simplest possible calculation there is: 1 + 1. The fact that we happen to be adding thirds, as opposed to apples, marbles or anything else, is beside the point. Or so you would think, if you're comfortable with fractions. But a lot of children are not ...

I can remember, at 9 years old, sitting in my classroom, when a girl called Sarah was asked to add together two fractions that had the same denominator. Because they had the same number on the bottom they were the same kind of thing, so it should have been easy, but she stared forlornly at the board and said that she didn't know. Helpfully, the teacher pointed out that it was just like adding oranges – simple arithmetic: 'Don't think about it as two-sixths plus one-sixth. Think of it as two plus one.'

Sarah shook her head as the rest of us gasped or rolled our eyes. 'I don't know,' she croaked. Of course, this tells us more about the psychology of being put under pressure in front of a group of people than anything else. But, as well as that, the story shows something quite different: all too often children don't really know what fractions *are*.

Most teachers introduce the concept of fractions through the idea of sharing. Children have a very acute sense of fairness and fair sharing makes a lot of sense to them. So, if we divide a cake or a pizza between six people we want to make sure that each piece is the same size. These pieces are called sixths. That's fine. Kids get that, so teachers are probably starting in the right place. But the easy mistake to make is to abandon that approach from then on, because we think the kids have understood the concept of fractions. Here's why …

The first problem is in the names: *whole, half, third, quarter, fifth, sixth, seventh* … These words are known to children already – to come third in a race, for example – but they have a different meaning when they are the names of fractions, which is not so helpful. Also, *half* and *quarter* have their own special names (we don't say *second*, and in everyday life rarely say *fourth*), so the sequence isn't logical – and children *should* be looking for logical sequences in maths.

The next tricky area is in the notation. Fractions look weird and scary. Admittedly, the notation is logical enough, because the denominator (bottom bit) shows how you get the parts, and the numerator (top bit) shows how many of those parts there are. But this whole made-of-two-numbers-with-a-line-in-between doesn't sit easily with many people. What is this thing I am looking at, they wonder. A fraction is like a number, but isn't a number. Or not exactly.

So, what is a fraction? The most common definition I come across is 'a part of a whole'. That should be fine. Children understand what a bit, part, bite, chunk or piece of something is. With this existing concept, they can quickly move on to the idea of an *equal* part of something, which is how children are presented with fractions, as I said. The thing is, though, dividing a pizza equally between a certain number of people is an introduction to the *unit* fractions $^1/_2, ^1/_3, ^1/_4,$ $^1/_5$, etc. What about the others – $^3/_4, ^9/_{10}$, etc.? The leap between $^1/_4$ and $^3/_4$ is enormous. Why?

Well, $^1/_4$ is one part of something bigger. But $^3/_4$ … is that *three* parts of something or *one big* part of something? Perhaps both are true, and the difference might seem unimportant, but someone learning these things for the first time can get snagged on an issue like this. Imagine the pizza already cut up into four quarters. In that case, $^3/_4$ is three parts of something, like three pieces of pizza. Imagine, instead, that you started by cutting one quarter out and leave the rest of the pizza intact. Now $^3/_4$ is one big part of something. But the number has a 3 in it. And a 4. How does the combination of those two numbers represent the Pac-Man-shaped pizza with a bit missing that we now have? It's not that obvious. Show diagrams of both kinds of $^3/_4$ and not all children will comfortably assert that they are both three quarters.

Which leads me on to my next point. Children need to visualise fractions. When they think of ten they can call to mind what a group of ten objects look like, which gives them a handle on what ten is. When they think of a given fraction, they need to be able to visualise it just as easily.

It doesn't have to be accurate, just a basic idea. For example, try to visualise these fractions:

$^2/_7$    $^2/_{20}$    $^9/_{10}$    $^{11}/_{23}$

Now match the fractions to these diagrams:

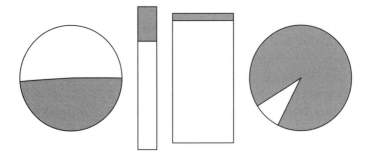

No doubt, you originally visualised them differently from the way I've sketched them, but I predict that you can quickly spot which is meant to be which. You might, for example, have quickly assessed whether the fraction was more or less than half, very small or very big. That is because fractions actually mean something to you.

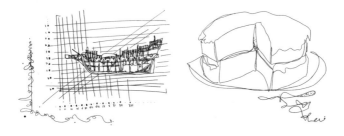

We need to make fractions mean something to the learner, and be a part of their understanding of number and quantity. Here is my way of attacking the problem.

# Things to Do 19

**Do** Put a number line on the floor.

Or you could lay it out on tables and all gather round. Just make sure there are some big spaces between the numbers. Children may have comments before you even start, just from what they have seen so far. Hear them out in case you can use them as a starting point. Before long, though, you are going to ask:

**Say** Is there anything in-between the numbers?

What are the points in-between the numbers? Are they numbers too?

How many are there between each number?

What can we call them (to tell them apart from each other)?

After discussing this with each other they might all say that there is nothing between the numbers. If so, move on to the example below. If they do come up with ideas for what is in-between, ask them *what kind of thing* is between, say, 1 and 2. Is it a number? Or something else?

**Do** Have the numbers represent age, so that someone starts at zero when they are born and moves along the line. Use a toy or similar figure as a counter to show someone moving along and getting older.

**Say** On your fifth birthday you are 5. After that do you go straight from 5 to 6? How old are you in-between? What happens during that year?

Sooner or later someone will come up with 'five and a half' as an in-between point (if they don't, put your counter in the right position to elicit it). Get the child to show on the number line where that is, if you haven't already.

**Do** Bring in another figure to represent another child. Put the two figures in slightly different places on the line, always between two whole numbers.

**Say** How can we say what the difference is in their age?

Someone will probably come up with the idea of splitting the year into months as a way of comparing the gaps between the ages of the two figures (and if no one does, bring in birthdays as a clue). This is perfect because now they are talking about fractions without realising it. At this point (or earlier if you had to), you could bring in the birth dates of the class. If we are comparing their ages, we will need our fractions (i.e. days, weeks, months) to say who is older or younger.

**Do** Write up all the ages of the children in the class in this form:

Lucy – 8 years and 4 months

Laila – 8 years and 7 months

**Say** How can I write these ages using just numbers, without words?

All they need to do is switch to using the word 'twelfth' and writing it: $/_{12}$. A twelfth is what you get when you divide something by twelve. We can use these to talk about what is in-between the whole numbers.

# Key Words 8: Known to Unknown

The best piece of advice I ever had on teaching was in Zimbabwe when I was a 19-year-old, enthusiastic but unskilled temporary teacher in a primary school. During breaks the male teachers would stand around in the open air and chat in one group, while the women stood or sat in another circle. One of the women, the headmaster's wife, was often knitting, but I never heard much of what the women were discussing. The men discussed either politics or teaching. One day my colleague, Mr Pfunde, told me, 'Teaching is taking people from the known to the unknown.'

It doesn't sound very profound or earth-shattering. But the bit people forget is that you have to *start with the known*. Learners have to link new ideas to ideas that are already in place, and too often we start by introducing the unknown and don't give enough thought to the links and comparisons we can make to what is known. Without these the new concept is floating in space, and even if you succeed in teaching children to operate that concept in a particular way, they fail to apply it in other contexts or make it part of their toolkit for solving problems.

The activity in this chapter (and the next one, too) is an example of 'the known to the unknown'. We are digging into their knowledge of time and dates – that is the known. The unknown here is the idea of the fraction.

# Half to You

What is knowledge? This is one of the oldest and toughest philosophical problems. Like all the old, tough problems, it is something we feel we instinctively know the answer to, but can't quite get a grip on verbally. That is to say: we know what knowledge is, but we just can't say.

Interestingly, some other languages have two words for our word 'know'. In French, you have *connaître* and *savoir*, while in German you have *kennen* and *wissen*. As you may know, the French *connaître* and the German *kennen* both have the same meaning, which is to be familiar (with something), as in, 'I know the house because I've been there.'

The other two verbs, *savoir* and *wissen*, are for knowing facts, as in, 'Do you know which house is hers?' And, if you think about it, the meanings are pretty different. I'm reminded of this difference sometimes in the classroom, when children hoist their hands high to tell me what they 'know' but they are really telling me what they know *of*; there is not much wider understanding or background to back it up.

Sometimes the children are aware of this themselves. They might say, 'I know that a triangle can be equilateral but I'm not sure what that means.' They should be praised for admitting this, for knowing the difference between a vague familiarity and real knowledge.

Which brings me back to the question of, what is knowledge? I won't attempt to define it now, but here is a distinction that I think is useful: there is a difference between 'doing' a topic and 'knowing' it. If you have a class who 'did' percentages last year, they will be able to spout various half-digested facts and connected vocabulary, but these misunderstandings will not be something you can build on. You will, in fact, have to go back and reteach the whole concept from first principles if you want to move them on from the 'level' they supposedly attained last year.

If the class really knows percentages then they will use them. Given a situation where it is appropriate or useful to talk in terms of per cent, they will think of doing it and achieve it with some success. They will be ready for the next step.

And this goes right to the heart of what education is – and isn't. To some people, it is the transfer of information and procedures from teacher to pupil. I show you how to multiply fractions, now you multiply fractions … you can now multiply fractions. We know stuff because we are told it.

Of course, it is hard to imagine a workable educational system *without* teachers ever telling people stuff. True, some progressive types have indeed tried hard to dispense with 'telling' altogether, and I'm grateful to them for trying, although whether their pupils were I don't know. For me, the thing is that you tell them when they are ready to be told. And the genius in teaching is *making people ready to be told*. You ignite their curiosity by making them smile or frown or swear, so that when you give them the missing piece to the puzzle, they feel a sense of satisfaction that something puzzling or incomprehensible is now complete.

One danger of a test-driven culture in schools is that children become familiar with a topic, know some isolated facts about it, but have not invested any of their own thinking in it. In fact, they don't know it at all when you come to a different kind of test. So, can the child reach for this concept spontaneously as a way to answer a question? Can they apply it competently and confidently?

If you have read this much of the book then I may well be preaching to the converted. So, the more relevant point at this stage is not that we must not teach narrowly but we must find out how not to teach narrowly.

Well, to create knowledge you have to 'waste' time. You have to explore the topic at the child's pace, answer her questions and get her to test out her learning to her own satisfaction – that is where the competence and the confidence will come from. Then, once the concept has sunk in, you can quit wasting time and step on the gas. Hopefully.

In this next session, you can explore what children already understand by the word 'half'. How do they apply it? Remember, the aim is to deepen your understanding of their thinking first, not correct things straightaway.

For example, let's say you asked this question: 'Can you have three halves of something?' A child may answer, 'Yes', 'No' or 'Don't know'. If a child says, 'Yes, you can have three halves of something,' then what can we conclude about their understanding of the concept of 'half'?

Not much – yet. It just depends what they are thinking. So we have to find out. Say, 'Show me', and get the child to show you physically, or pictorially, what they mean. If they divide a whole apple into three and announce that they have three halves you can then get them to compare an apple divided into two halves and an apple divided into three 'halves'. Gradually, you are unpacking their thinking and finding out how they conceive things.

Or perhaps the child shows you three halves by drawing or picking out one-and-a-half apples. In this case, the child is absolutely right that there are three halves of half something: three halves of apple.

If controversy breaks out in the classroom ('That's not three halves!', 'Yes, it is!') all the better. Teach the children to listen first, address what has been said and show respect with the words they use.

Knowledge is not data downloaded into our heads. It is more like a map of the world, subtly different from everyone else's, that we have a hand in building ourselves from the common data.

# Things to Do 20

This is another way of introducing fractions. We start with the fraction that pretty much all children know already – the half. When I say they know this fraction, what I mean is that they will know the word and use it quite confidently. In this activity, you will get to see how they currently understand it, and then you can build on, or refine, their existing understanding, before introducing other fractions.

**Do** Assemble a collection of different types of 'halves' and present them to the class.

Your collection should include things that are clearly not halved exactly and some that are. Here are some suggestions:

❖ Two halves of an apple.

❖ A pencil broken in half.

❖ Two halves of a picture.

❖ Some bisected shapes – triangle, rectangle, circle, etc.

❖ Something that is truly halved but not symmetrically – for example, a sheet of squared paper cut so that each piece has the same number of squares on it but is not the same shape as the other half.

Scatter them across the floor. Children will be keen to match them up.

**Say** Someone show me a half.

See what they bring you, and what they say about what other people have brought. You may need to prompt …

**Say** Are all these things halves?

The first step might be to separate true halves from false ones (i.e. elicit that halves must be equal). This is not obvious. If you divide a muffin in half it does kind of make sense to say, 'You can have the bigger half,' even though, mathematically speaking, 'bigger half' is a contradiction. Try not to tell the children they are wrong to see two unequal parts of something as halves; it is more that in maths halves must be equal.

Then move on.

**Do** Pick up one of the divided items, say an apple, and reunite the parts in your hand.

**Say** Do I have an apple in my hand?

With a bit of luck there will be real controversy among the kids as to whether you have two halves of an apple, or one apple, or both. Let it run for a minute or two if it engages the children. What you are seeing is them getting to the very nub of the conceptual problem: 1 is 1, but it is also $1/1$, $2/2$, $3/3$, $4/4$, etc., if we choose to see it that way or split it that way. To become adept at maths, you need to see each number as made up of various other small ones and an ingredient in lots of bigger ones. It is a fluid way of thinking.

The next stage is to pick up one of the halves and halve that.

**Say** What have I made now?

If they say, 'half a half' then tell them that is exactly right. Then ask if a half of a half has a name. The children may well be able to answer 'a quarter'. But dig down a bit if they do: ask how many quarters of the apple there are (remember at this stage you have three pieces – two quarters and an uncut half). If they stick with the answer 'two', again, that is perfectly correct in one way, but get them to look at that uncut half. There are two quarters in that half, if you think of it that way. Just as every whole is $^4/_4$ and $^{10}/_{10}$, every half is $^2/_4$ and $^5/_{10}$, etc.

Take your time over these issues. Let the children stumble and air all their misconceptions. Take those misconceived ideas as seriously as you can, if they are meant sincerely. The children must only give up their mistaken belief because they have seen it is mistaken, not because they take your word for it. Otherwise, their new 'correct' belief is not a solid foundation for further knowledge to grow on.

So, I am saying that, conceptually, the natural first step in understanding equivalent fractions – perhaps all fractions – is to understand that:

$$1 = ^1/_1 = ^2/_2 = ^3/_3 = ^4/_4 \ldots = ^{100}/_{100} \ldots$$

I would get the children to look at the apple, and visualise it cut into two equal pieces, then ten equal pieces, then twelve and so on.

If they are all happy with this then you can start working towards a wall display of equivalent fractions. Give them each (or in pairs, whatever) some shapes cut out of squared paper. Use shapes they are familiar with and have some new ones as back-up in case they find it easy. Then get them to cut these shapes into different fractions. After they have done a couple, tell them to try to think of ways that no one else will have thought of.

There are different ways you can run the activity. For example, you could give each person or group the same variety of shapes, and see which different kinds of fractions they cut them into. Or you could give each group multiple copies of the same shape (e.g. eight different 12 x 5 rectangles) and ask them to cut each one into a different set of fractions (e.g. fifths, twelfths, tenths, sixths, etc.).

Whichever way you approach it, the aim should be to show that a whole can be divided into fractions in lots of ways, and that lots of fractions are the same size as each other (i.e. equivalent).

# Things to Say 10

*'Show me!'*

A child has just said something you don't quite understand. They may be on to something, they may not. You want to get them to explain, but at the same time you don't want to get them to explain, because they might be going off on some awful tangent.

Alternatively, a child has just come up with something brilliant – the moment of insight that will illuminate the whole class.

In both cases by saying, 'Show me!' you make the right next move. Here are some reasons why:

- ❖ It is a call to action.
- ❖ By stepping back, you centre everything on the learner and her perspective.

- ❖ The class gets the chance to see someone's thinking, as well as hear it.

- ❖ The teacher is expressing interest, without confirming what is correct.

- ❖ You give the class confidence to air their ideas and put them into action.

Obviously, this simple phrase is appropriate for when we are dealing with pictures and physical objects. And that is no coincidence, because if the class are using these as aids to thinking then you will probably keep them on task and interested for longer.

If you are not saying 'Show me!', you end up saying things like:

OK, I'm not sure if I quite understand. Could you explain again?

Thank you, that's interesting. Has anyone else got an idea?

Is that the same as Kevin said? I think so …

I am not saying these are terrible things to say, but they are not as dynamic.

# Action Fractions

Why bother? I mean, do we really *need* fractions? After all, when was the last time you added, subtracted, multiplied or divided some fractions? Is there any use for them in everyday life?

You could argue that there isn't. Take this example of a problem:

---

A woman inherits a piece of land of 900 square metres along with her four sisters. So she stands to inherit a fifth of it. However, she is completing a messy divorce where her husband has been awarded two-thirds of everything they jointly own, including this land. How many square metres will he get?

---

I would start by calculating her share of the inheritance: so I need to divide 900 (the total area of land) by 5 (the number of siblings) which is 180. That's the woman's own share taken care of. Now, the woman's share is jointly owned by the husband, so she has to give two-thirds of that share to her grasping ex-spouse, so I'd divide that share by 3 – divide 180 by 3, that's 60 – and double that to get two-thirds: 120.

There are other ways of doing it. But I don't think many people would work with the fractions in the abstract and start by wondering what two-thirds of a fifth is. (If you are one of them, that probably says great things about you, but you are going to have to accept you are in the minority.)

In fact, all I did was divide things and multiply at the end. So I wasn't dealing with fractions, as such, at all. I could do all the calculations without knowing the word for 'fraction' or writing one. Of course, you could *call* some of those figures fractions (e.g. you could say that the 180 was a fifth of the total land). The thing is, though, when I multiply I could just as easily *call* the numbers that result 'products' – that is what they are – but I wouldn't, and I don't.

So, come on then, what are fractions *for*? What would life be like if we didn't have them?

Well, the Romans seemed to do alright, and they didn't really have fractions: they used twelfths. Twelfths have the advantage of being easy to work with: a quarter and a third can be easily compared, added or subtracted – being three- and four-twelfths, respectively. And we should remember that the Romans were brilliant engineers – they built aqueducts that brought fresh water many miles to feed cities (the longest, to Constantinople, was 250 km), and those aqueducts had to run downhill by tiny amounts all the way. So, twelfths seem to have done them fine. The Egyptians only used unit fractions and a few obvious ones like $^2/_3$ and $^3/_4$. They got the pyramids up just fine.

But, of course, we can't just decide not to bother with fractions. They are in the maths curriculum and aren't going away any time soon. Any argument about whether we should or shouldn't teach them is hypothetical, as we have to teach them anyway.

True. But whatever we are given to teach in the curriculum, in order to teach it we have got to find reasons why it is worth teaching and communicate that through what we do. Otherwise, we will just end up as zombies teaching dead knowledge. Neither you nor I is at that stage yet.

To explain why I think fractions are necessary, I should first admit that I slightly cheated in saying we don't use fractions much. Because it is easy to forget that two kinds of fraction are used all the time: the decimal fraction and the percentage. Statistics and comparisons are full of them. If that is the case, someone might ask, why not just teach those?

I think the answer to this is that we can't – I don't think, anyway – really teach decimals and percentages without first teaching numerator/ denominator fractions. The concept of tenths, hundreds and so on seems hard to comprehend without knowing fractions. If we could find a way of teaching them without common fractions, I think we should seriously consider it. But until then we will have to get stuck in.

Fractions are ways of saying how much of a thing we are talking about. Sometimes it is not enough to say 'a big bit', 'a tiny bit', 'a lot' and so on. We need to be more precise. So, we divide the whole thing into equal parts and say how many of those parts we have. It doesn't matter too much whether we divide into tenths, twentieths, thirty-thirds or thousandths, as long as what we are saying is understood. At the end of the day we have to agree on what we are talking about.

And one thing we always need to agree about, to avoid trouble, is money.

# Things to Do 21

The example below is a scenario designed to get kids thinking about fractions in a motivating, immediate way. Like the case I gave above, you can solve the problems here without using fractions as such – just use the four basic operations – but the numerator/denominator notation does help when we are saying, 'One child should get two-fifths of the total,' rather than, '… should get the total divided by five and multiplied by two.'

In a way, we are trying to recreate the process where human beings first started using fractions – because they *saw they were there* and needed to call them something.

**Do** Tell this story. You can read it out or tell it in your own words, obviously.

Two teenage brothers live in the same town as Wayne Rooney. One day they are trying to make some extra money by washing cars. They knock on Wayne Rooney's door and Wayne says they can wash his cars. He has five cars. The older brother is faster and he washes three. The younger brother only washes two. When they have finished Wayne Rooney gives them £10.

**Say** How can they split the money fairly?

I'd recommend you use pictures or toy cars, figures, and real/fake money. Let them wrestle with this for a bit, apply and test their own strategies, and come to conclusions about whether they can do it, and how. Then adjust the scenario to create different calculations. The idea is not to fire all these next questions at them, but to select a question they can answer and then make the questions more complex, step by step. Rearrange visuals and coins/tokens to show how the situation changes and try to give realistic reasons for the changes.

- ❖ The next day one of the cars has gone to the garage and Coleen has driven another to the gym, so they only have three to wash. How can they split the work and the £10?

- ❖ The next day Wayne has bought one more car. If they did three cars each, how would they split £10?

- ❖ Wayne is so pleased with the work that he invites Robin van Persie to bring two of his cars to be washed … Because they're getting tired they invite a friend to help them. What if one of the boys was going to wash one car but suddenly had to go home, without doing anything? The oldest boy says, 'I'll do your car.' How many does the oldest boy do? How many do the others each do? How would they split £10?

If you think the class are going to struggle, then start with the easiest way of presenting it: five cars, five boys, each boy does one car and they split £5. There is no harm in starting here even if you think it will be easy for them. You just speed through to more complex examples.

You can alter the variables to make it more fiendish: one boy only does half a car, two boys do the last car together, Wayne only has a £20 note and lets them keep the change, the cars are different sizes … and so on. And see if you can get the children to devise variations themselves.

Remember that as long as each of the boys in the scenario is happy with his share of the money, then the problem is solved.

I worked on this problem with a mixed ability Year 4 group for several sessions spread over several weeks, and each week the class could recall the outline of the scenario easily and I just fiddled with one of the variables to make it new, and to take them on to new ground.

# Key Word 9: Differentiation

There is a lot of controversy and research about whether dividing children into groups according to their ability – streaming – is a good thing or a bad thing to do. There is evidence that it improves results, and contradictory evidence that it makes results worse. As ever, the situation is too complex to say that streaming works or doesn't. Japan and Finland don't stream much, if at all, and have some of the best maths results in the world. Norway and Hong Kong stream more, and also have some of the best results. The US, UK, Canada, Australia, Ireland and New Zealand stream quite a lot and have some of the worst results relative to the wealth of those nations.

What definitely seems to make a difference (and it is no surprise) is the way the teacher adapts the content of the lesson to the individuals in the classroom – whether they are streamed or mixed ability, you still have a range of abilities and personalities. This is usually called 'differentiation'.

When children are sharing their thoughts out loud, it is a great opportunity for insights to spread among the class. Kids pick up ideas very quickly from each other, partly because the language children use is sometimes the most accessible to other children.

Furthermore, in this activity, you can see that the class as a whole can progress to a certain level, and then some children can take it further. It is important that the teacher elicits from the children themselves ways to extend the task. What you want to do is foster an atmosphere where, once they have completed the initial task, children explore around and beyond it. In this way they are doing the differentiation themselves. They can do this by asking themselves these sorts of questions:

* What if there were more?

* What if there were less?

* Does it work the other way round?

* Is it the same as … ?

* What is the next step?

The questions can come from the teacher to begin with, but soon the children should take over and adopt this attitude themselves. They will do so if their first efforts at it are encouraged. All human knowledge stems from people thinking, 'I wonder if … ?', and then following their curiosity. That is the best way to get your children enlarging their horizons, stretching their capacity and amusing themselves.

# Go Compare

So, they've done fractions. Tick. And decimals. Tick. And place value below the decimal point for tenths and hundredths. Tick. Next on the list is percentages. Shouldn't be too tricky. The children need to be able to convert fractions and decimals into percentages, and back again, and work out percentages of various amounts. There is your objective, right there. Job done?

This is one of those times when we have to ask ourselves what the objective of the objective is. In other words, why are we doing this? Here is my exploration of why we learn percentages.

Start by trying to put these fractions in order of size:

$$\frac{3}{8} \quad \frac{2}{7} \quad \frac{12}{33} \quad \frac{1}{4} \quad \frac{8}{25}$$

Actually, don't. I'm just trying to prove the point that it would be fiddly, however you tried to go about it. Now, can you put these fractions in order?

$$\frac{38}{100} \quad \frac{29}{100} \quad \frac{36}{100} \quad \frac{25}{100} \quad \frac{32}{100}$$

Not difficult at all. In fact, there is barely anything to do. What you may or may not have noticed is that the second set of fractions is the same as the first – almost. What I did was put the first set into the calculator and divided (3 divided by 8 and so on). Then I rounded the answer to the nearest hundredth. And that, of course, is how we create percentages:

---

38%    29%    36%    25%    32%

---

Amounts expressed in this way can be compared directly and their relation to each other is instantly apparent. This makes them a useful tool for analysis, and for converting numerical information into a form from which we can draw conclusions. And since the arrival of the calculator in the 1970s (and, boy, were they impressive back then) percentages are very easy to produce.

However, percentages do have one important shortcoming: they are not often accurate. The percentages I give above are not all identical to the fractions I started with. The last two in the row are identical to their counterparts, because 4 and 25 are factors of 100, so I can convert quarters and twenty-fifths into hundreds perfectly. The other three percentages are rounded up and so, ultimately, false. Three-eighths isn't actually thirty-eight hundredths.

It all comes down to the level of accuracy we need versus the level of convenience we want. And the interesting thing is that, most of time, for most of what we get up to, percentages are accurate enough. Differences of less than 1% are usually negligible. Admittedly, if you are dealing with a million dollars then 0.6% is $6,000, which is a lot. Then again, perhaps it's not a lot to someone who has got a million. You and I will probably never know.

So, we have answered the question of what percentages are for: they are for comparing amounts easily. The other big question is what actually *are* they? Luckily, the answer is very simple, and we have already covered it. Percentages are hundredths. There is no difference between 20% and $^{20}/_{100}$ except how we say and write them. That is why it is odd that on the couple of occasions I've worked with children who are supposed to know percentages, they can rarely tell me that percentages are fractions. Even when I put it to the individual that fractions is all they are, there is a lot of uncertainty.

# Things to Do 22

If the point of percentages is comparison, you need some things to compare – some data. You have two options: use existing data or create some. The latter choice is not always necessary or desirable. If you are looking to link this work to science then, yes, generating your own data is a great thing to do. If you are linking it to history then researching data that already exists is just as important.

The best subject of data is often the children themselves. They are automatically interested in information about themselves. Here are some everyday facts that will interest most groups:

- ❖ Their gender.
- ❖ Their months of birth.
- ❖ The distance of their home from school.
- ❖ Their heights and weights (though this can be sensitive too, so shoe size or hand span might be preferable).
- ❖ Pet ownership.
- ❖ Football team supported.
- ❖ Number and gender of siblings.
- ❖ Their countries of origin, if varied.

**Do** Give each child a sticky note, postcard, index card or piece of paper of that size on which they write their own name.[1]

They can decorate it a bit too. This card will now represent that child. You then give the class a reference or measure – for example, gender:

**Say** How can we use these cards to show how many of the class are boys and how many are girls? Are we mostly boys or mostly girls?

The children should converge on the idea of displaying the cards in two groups – girls and boys.

You may find that it feels pointless to be converting this data to percentages at this point. And so far it is, and that is because we are not comparing yet, and I have been arguing that comparing is what percentages are all about.

**Do** Show a similar set of data from another group (e.g. taken from the school records).

**Say** Are they mostly boys or mostly girls?

For your comparison group, try to use one that has a different total number of members. This is because if you compare, say, the gender balance in two groups of thirty it is quite easy, because you will be working with a common denominator: $^{14}/_{30}$ of one class and $^{17}/_{30}$ of the other are boys. Whereas if you have a class of $^{14}/_{30}$ boys and another of $^{15}/_{33}$ it is going to be hard to compare. This is where percentages come into their own – converting a tricky task into an easy one.

If the class has never been introduced to percentages, you now have them at the perfect point – because there is an obvious purpose to them here. But, first of all, reinforce the concept that *comparing with*

---

[1] The advantage of sticky notes is that you can put them on a vertical surface and move them around. The whole thing can be done equally well, though, with postcard-size slips of paper on the ground.

*common denominators is easy.* From there you can move on to a common denominator that we can use for everything, whenever we want. This has been chosen to be hundredths. And we have a special way to write them – with the one being slanted over between two zeroes to show that this is a special type of hundredth.

# Key Word 10: Discovery

What I mean by discovery is that the children's experience comes before their instruction. So, rather than start by outlining a concept on the board, and naming it, I would usually try to launch straight in to a story, question, experiment or other stimulus – and get the children to stumble across the point of the lesson. After they have discovered and discussed the idea, then I will step in and put a name to it. Most of the Things to Do in this book take that form.

The reason I prefer this format is that, as I have emphasised before, it creates a need for the knowledge. The children are open to the idea because they have just been dealing with it. They have seen it emerge from the activity.

The opposite structure is to work down a kind of conceptual pyramid:

* At the top is the *name* of the concept – for example, percentages.

* Then comes the *notation* – 10%, 25%, 42%.

* Then come the *definitions* or explanations – a way of showing amounts in hundredths.

* Then come the *abstract examples* – 44% of people in the UK own a car.

❖ Then – if you're lucky – come the *practical problems*.

If you are dealing with conceptual classifications this is perfectly logical – start with the most general and abstract, and work down to the specific. But, for me, it fails in educational terms because it doesn't put the purpose of this mathematical practice at the heart of the lesson. This is why I start from the bottom of that pyramid and work up.

The moral of the story, for me, is: don't name it till they know it. In other words, name something they have already come to understand and see the sense of.

# Your Average Number

There is no such thing as the average family. In 2006, the average number of children per family in the UK was 1.8. Of course, you don't need me to tell you that no family in Britain had 1.8 kids. That's impossible. Therefore, the conclusion is that not one single family in the entire country is average.

As always, it depends what you mean by 'average'. In conversation, we tend to use the word as a synonym for 'normal' or 'typical'. Whereas when we calculate an average for something mathematically we usually add up all the items in the group and then divide by the number of members in the group, which is how you get the 1.8 figure. As you may be aware, the exact name for this type of average is the *mean*.

People who are careful about statistics often specify when they are talking about the mean number of something, because there are other ways of calculating averages, as you probably know. Actually, there is nothing to stop you working out your own kind of average. That is because the reason we use averages is so that one figure can represent a whole group, so that means it is up to us to decide what figure does that job best.

For example, if I wanted to compare the earnings of carpenters and plumbers it would be silly to compare the total amount earned by all carpenters with the total amount earned by all plumbers. The carpenters' total might be higher just because there were more carpenters around than plumbers. And just looking at the highest or lowest-paid people in each job could be misleading too. So, calculating a mean makes sense. However, this way of doing it (dividing the total amount of earnings by the total number of earners) is arbitrary – there is no mathematical truth or rightness to it. It just makes practical sense, because in so many cases it gives an informative result.

Now, let's look at two other well-known and commonly taught ways of reaching an average: the *mode* and the *median*. The mode is the most common number (I remember this by thinking that 'mode' is French for 'fashion'), so in the set 1, 2, 4, 4, 5, 5, 7, 7, 8, 8, 8, 9, 9 the mode is 8 because it occurs more than any number. The median (which sort of sounds like 'middle one' to me) is the number in the middle when all the numbers are written in order of size. So, in the set 4, 8, 8, 9, 11, 12, 12 it would be 9 because it is the fourth number of seven and has three numbers either side of it.

And – here comes the important bit – the reason we need different types of average is because we have different types of groups, and different interests in them. Now, in some groups, such as the one below, the mode, median and mean would all come out the same:

---

100, 150, 200, 200, 200, 250, 300

---

As you can see, 200 is the middle number, the most frequently occurring number and the number you get if you add them all up and divide by seven (the number of group members). So, 200 is the median, mode and mean.

But most groups, especially of real data, are much more messy than that. Going back to our UK example of 1.8 kids per family, I would bet this mean average is fairly representative of the group. For a start, the mode is almost certainly two (lots of families have two nippers) and most families will have a number of kids very close to 1.8 (maybe 90% have between zero and three). So, yes, the mean of 1.8 is helpful.

But now let's take the example of a community where the rising middle class have small families because they pay a lot for education and want to put all their resources into ensuring the success of just one or two offspring. At the same time, there are poor families who either lack the access to birth control or feel that large families will help to look after them in their old age. So, you could have a mean average of four children per family across the whole community, when hardly anyone actually had four. Instead, 90% of the population had less than two or more than five. In this case, the mean wouldn't tell you much about the real situation.

These differences are really important when it comes to statistics about income and wealth. Don't trust the UK statistics I use here too much – let's just take them as possibly true for our current purposes:

Mean average income (full-time workers) = £28,600

If you add up all the money earned by everyone and divide by the number of earners this is what you get.

Median average income (full-time workers) = £23,764

This is the middle person in the list – 17% less than the mean. So, if that were your salary, then half the population would earn more than you and half would earn less.

The mode is … no one seems sure! The research, if it exists, is not easy to track down and the mode is rarely mentioned in discussions. But it is interesting because many people estimate that the income bracket with the most people in it is the minimum wage. For a full-timer that comes out at about £10,000–11,000 per year. Barely a third of the mean.

So, which of these averages would make you 'normal'? Well, that is a big question. The reason that the mean is the highest is that there is a tiny minority of people earning between ten and a hundred times what the rest of us get. Although it is a minority, their income is so high that it brings the national average up.

As I said, there are rival versions of these figures and I haven't allowed for tax and benefits, which obviously narrow the gap. Neither have I gone into disparities of age, gender and region – which are all huge. The point I am making is that in a complex group with unequal distribution, using an 'average' figure may be more likely to conceal than convey the truth. It can be misused by people on either side of an argument.

# Things to Do 23

We are going to put the class in a situation where they need to find some kind of average in order to make a decision. You could run the whole thing without ever referring to averages, or you could use it as a way of introducing the idea for the first time. Or you could use it to revise the concept. The depth at which you discuss and analyse will depend on their starting point and what you want to achieve.

It is all about spending money – pocket money, as it used to be called. And the scenario is: Leonardo, a child of … (whatever age your class is) is arguing with his parents about how much money he should get to spend.

**Say** Should children be given money of their own to spend? Why/ Why not? How much?

You want to move the discussion on just at the point that it starts to warm up, not try to run through to a conclusion. Once the debate is animated, extend the question:

**Say** How should the parents and the child resolve the disagreement?

Listen to a few ideas on this before describing our scenario:

Leonardo wants to get a weekly allowance from his mum. His mum eventually says: 'We will give you the same as the people in your class at school get.' The problem is that different children in Leonardo's class get different amounts.

**Do** Display the figures below:

| | |
|---|---|
| Jade | £4 |
| Coleen | £12 |
| Maggie | £2 |
| Polly | £7.50 |
| Neela | £3 |
| Louise | 0 |
| Emily | £2 |
| Tara | £8 |
| Sonia | £2 |
| Miriam | £3.50 |
| Cameron | £2 |
| Eddy | 0 |
| Troy | £1 |
| Tyler | £10 |
| Sami | £3.50 |
| William | £2 |
| Neil | £15 |
| Felix | £7 |

Bruno    £10

Robert   £5.50

---

You could, of course, gather your own authentic data, which would usually be my aim. The immediate problem in this case, however, is that the real disparities in cash will be a sensitive subject. Children may get upset or ashamed, and they will almost certainly lie. So fake data is probably best. I realise mine may not be realistic for your age group or part of town, so scale it up or down, but I recommend you multiply or divide *all* of them by the same amount, because I've arranged them so that the mean, mode and median come out different.

Now set them the puzzle:

**Say** How should the parents use this information to choose an amount for Leonardo?

In discussion, some children may try to reverse the decision of the parents and avoid using this information. Gently remind them what we are thinking about: '*If they do* decide to use it, how should they do it?'

Make a note of each suggested way of doing it. Examples might be: it should be the same as the child's best friend; it should be the same as the highest; you should choose one in the middle. If their ideas aren't very mathematical keep reminding them: are you really using this information? Will this method make Leonardo's spending money the same as the children in his class? Once you have a few, get the children to compare them. Then, if it hasn't happened already, get the class to select one method, try it out and see what number it produces. Do the same for the others.

Some of the methods they come up with may approximate to mode, median and mean. Of course, you have the option of introducing these three yourself, and naming them, but don't bring them in as 'the answers'. They are conventional practice not proven truths. I would recommend using the mouthpiece of other characters in the story (uncle, teacher, classmate) to avoid investing them with too much authority.

As I said, I've fixed those cash figures to get a fruitful outcome: the mode is £2, the median is £3.50 and the mean is £5. You could argue that each of these is or isn't fair to use as a benchmark for Leonardo's allowance. One class thought of taking the mean of the three different kinds of average, and I thought that was about right myself. But it is all about the reasoning with this task, not the final conclusion. If children honestly feel that Leonardo should get more than £10 if a few others do, there is no need to persuade them otherwise, unless that is a specific educational aim of yours.

The children may ask you if the numbers are made up. You can just say you got them from a book and you aren't sure if they are real or not. You don't want any families calling up the school and saying their teacher told the children they are entitled as human beings to receive £10 a week from whoever has the job of looking after them.

# Key Word 11: The Imaginary Disagreer

This technique is taken from Peter Worley's *The If Machine* and is an excellent way to redirect a discussion without stamping your own personality on it.[1]

There are times when you want to introduce a point of view that none of the class have spotted so far. If you introduce it as your own point of view, it will tend to stifle debate. Children often believe that

---

1 Worley, *The If Machine*, p. 32.

the teacher's idea must be 'right', and won't challenge it even if, at heart, they disagree.

So, one way to feed other ideas into the debate is to say, 'What if someone said … ?' Or, better still, say that a child in another class expressed the view or some other person you know. This is a lie, but I think it's OK!

This is a vital tool because, while you want the discussion to be free and for the children to follow their own ideas, some discussions can get stuck and need restarting. The children don't need to be persuaded by the Imaginary Disagreer's opinion; they just need to engage with it so that the conversation moves on to new, more fertile, ground.

# Our Street

I think it was a Saturday. Dad said we were going out for a look around and to take some pictures. 'On the lookout for shapes and patterns,' he said. Whenever I noticed one, I was to tell him and he would photograph it.

This sort of thing happened occasionally. My dad was a scientist originally but had decided to go into teacher training shortly before I was born. In those days, teachers and schools designed their own curriculums so coming up with exciting, imaginative topics for children to study was part of their job. Dad's was to open their eyes to those opportunities, and he practised at home.

From time to time he would arrive with bulbs and crocodile clips, or a machine the size of a small suitcase called a computer, or a pair of wellies and a shrimp net, and set up some kind of demo or field experiment. On one occasion, he tied a rope to the garden gate and sent waves down the rope by jerking the other end. The waves ran neatly down to the gate and he demonstrated wave frequency, length and amplitude by starting the waves in different ways. It was easy to understand because you could see it.

What we were doing this time wasn't clear. But it was always nice to get a bit of one-to-one time with him, so I put on my coat and we headed out. At first, he had to point everything out to me: look at that paving stone, what about that gatepost, can you see the triangle on the gable? The good news was that I soon realised that we could take as many pictures as we wanted (a big deal in those days, when you had to pay for rolls of film and developing). Soon I noticed that some wrought-iron gates (still big in the 1970s) had all kinds of patterns in them. One very common motif was a curly one that is called a scroll.

Gradually, I started to see more. Geometry was emerging from the very fabric of the environment. Everything built by humans seemed to be built out of shapes – you just had to look: hexagons in a wire fence, tessellating L-shapes on a manhole cover, the face of the wall was an arrangement of oblongs and squares. It was like a treasure hunt.

There is so much in the world around us that we just don't see as we walk down the street. There are probably good evolutionary reasons why. Hazards abound as we cross the road. Things to eat pass under our noses. Other humans bear down on us. Anyone whose reaction to a complex environment was, 'Oh look, a rhombus!', would probably not have lived too long in any period of history. But once you have made the decision to search your visual field for shapes, you won't be disappointed.

When I started on this book, I thought I would recreate the trip that my dad took me on but in the city where I now live. At random, I got my phone out on the unremarkable walk down a main road from the station to one of the schools I visit. I got twenty-four pictures. These are my best attempts to describe some of the things I spotted:

❖ Gatepost with a square surrounded by four oblongs to make a larger square.

❖ A chevron arrangement of oblongs in the pavement.

❖ Four different types of pattern in layers in one wall.

❖ A gate with mirror-image spirals and a kind of concave rhombus.

❖ Back-to-back question mark scrolls in a gate.

❖ Fire hydrant with rows of equilateral triangles in alternating orientation.

- ❖ A manhole covered in z-shaped octagons.

- ❖ A buttress with alternating brick layers, crowned with a triangle formed from five different brick shapes.

- ❖ Paving studded with low domes to help blind people locate crossings.

- ❖ A school sign in the shape of an oblong with a crest formed from three curves.

- ❖ Obelisk-shaped railing spikes formed from a truncated, elongated pyramid crowned with an equilateral pyramid.

- ❖ A window wall divided into twenty squares, eight blacked out and six with smaller inset squares to form an opening window.

- ❖ Gold circles alternately spaced in railings in front of a chain fence of squares tilted at 45°.

- ❖ Railings topped with gold fan-shapes mounted on globes.

Remember this was urban Britain, not Florence, yet still there were all kinds of details that were decided on by someone, drawn up and then executed. Some of them improve the visual quality of our environment. Some seem plain weird. But that is half the fun.

A trip like this, for children, is an antidote to what you might call the School vs. World problem, which is that kids spend half their time learning things at school, and half their time learning a whole lot of other supposedly unrelated things in the world outside. And most of the time they won't connect them up, especially if the people in their families are not inclined to make those connections for them. A good example of this is money. Kids spend a fair bit of time thinking about money and what they could buy with it. But when presented with money-based maths problems they look as bewildered as if they were expecting to barter for food with flint tools.

So, what we are doing in this chapter is showing that the patterns we study in the abstract are used in construction and decoration all the time. I recommend using both IT and old-fashioned pencils to make shapes and designs.

# Things to Do 24

**Do** Go for a walk with the class around some of the local streets. Take photos, especially close-ups.

It can be hard to organise taking the whole class outside the school gates, and you may be able to do the whole thing without leaving the premises, especially if your school has some older buildings, as they tend to feature all sorts of architectural decoration.

But I think staying on school grounds has to be plan B, because the very point of this is to get children to open their eyes *outside* school. If you are in a built-up area, think of a short route that takes you past different types and ages of building, a main road and some back roads. This is the best way to get a mix – and a few hundred metres may well do it. While you do need to recce the route in advance to make sure there are things to find, try not to find it all yourself first, as you want the children to point out things that you hadn't spotted.

You can introduce a competitive element by asking who will be the first to find a certain shape. You could print off a set of shapes and ask children to look out for them, then add some of their own.

Try to give some children the chance to use the camera, and allow them to practise transferring the images from the device to a computer.

**Do**  Go back and print off the photos.

Black-and-white photos would keep the costs down, but try to make sure they are of sufficient size and resolution. You could involve the class in choosing which of the pictures to select by describing the task ahead of them. Encourage them to choose some easy ones and some ambitious ones. Ideally, each child would have a set of images on a single sheet of A4 to work from.

**Say**  Can you draw a shape from one of these pictures – exactly?

They will need a ruler, pencil and squared paper. They may decide to draw the same shape either to scale or to actual size. A compass will be needed for any shapes with curves. Before they even start, emphasise that they will need lots of goes. They should be ready with the eraser and it may take a while.

As with any task, you have a choice of extensions from here. If all the children attempt to draw the same figure and share ideas on how they can achieve it, you will put more emphasis on technique, strategy and theory. If they are allowed to try different ones it becomes more of an art lesson, and a chance to go their own way. Defining or describing the shapes and their conglomerations may also be a part of your lesson – perhaps leading to something like my slightly comical attempts above (see pages 212–213)!

Kids could also think a bit about why a feature they have identified is there: is it functional? Is it decorative? Both? If decorative, do they agree that it looks nice? Does that matter?

You can make it more creative by allowing children to design their own wall, gatepost, fence – or even manhole cover!

# Things to Say 11

*'Why can't you do it?'*

'Ah, Miss, I can't DO it!' is the often heard cry of frustration. The pencil gets thrown down and so do the eyes, glaring into the corner of the room.

This is pretty discouraging, of course. When this happens to me I feel that I have pitched the lesson too high, bungled the instructions, paced or ordered or conceived or begun the lesson wrongly. Or all of the above. But it doesn't necessarily matter if I have. All is not lost – yet.

If a child is cross because they can't do it, that is a good thing in one way. They wouldn't be cross if they didn't *want* to do it. So, the first thing I tell myself is that this is better than the child being unable to do it and not caring. What I should hear in that infuriated exclamation is not a rejection of the task but a thwarted desire to succeed.

The second thing I tell myself is that even if the frustration is my fault, inasmuch as I miscalculated what I needed to do for this child, if I can make the most of this opportunity to put it right then this could be a win-win because the person I am trying to help will understand what actually motivates me – which is helping them to progress.

Third, I give the child a chance to express what they are feeling and explain why they are having problems. I ask more questions to draw out what they are thinking. If what they say is in any way reasonable, I accept it. Do all this before you wade in with a solution.

Sometimes the child will go sullen, shrug and clam up. OK, this might be a refusal to cooperate. Or it may be a lack of confidence – in themselves and in adults. I remember in one of my school reports an art teacher wrote: 'I can't tell if he thinks the work is beneath him or beyond him.' 'Beyond him' was the answer, but my self-image would not have allowed me to say that explicitly. I wanted the teacher to be able to tell without me admitting it. With adolescents, or children who have fast-tracked into adolescent behaviour in their younger years, a lack of confidence simply has to be covered up with bravado or feigned indifference. You have the right to take that at face value if you prefer, but you can also choose to try to find what lies underneath, even if it costs you a rejection.

# Sets Education[1]

What do your numbers look like? What colour are they? What pattern are they arranged in? What temperament does each one have?

These might sound like crazy questions, but people who can perform amazing feats of calculation often talk about numbers in those terms. You might think that the great human calculators see numbers in purely logical ways, but it seems that the opposite is often the case – and that might be a clue to their abilities. Even lesser mortals can benefit from a vivid sensory appreciation of numbers.

I used to imagine that the numbers one to twenty were like a tower of blocks. Then each set of ten after that would be arranged in a circle, perhaps reflecting the fact that the new numbers work through zero to nine and then start again in the units column. Unfortunately, the circles don't really make sense because then twenty-nine would be at the end of the circle, and so next to twenty instead of thirty. Maybe that is why arithmetic isn't my strong point.

Not long ago, a friend told me that she was hopeless with maths – even basic adding of two-digit numbers. But she had recently had an interesting conversation with a very old school-friend after they had got back in touch. Her friend made a comment that really surprised her: 'It all changed for you and me when they bumped us up a year and put us into separate classes.'

This brought a whole lot of my friend's memories flooding back. She suddenly remembered the days before being 'bumped up', when she

---

1  Some of these activities are adapted from the idea of 'People Maths' in Ollerton, M. (2003) *Getting the Buggers to Add Up.*

would go up to the teacher's desk time after time for more work, more tests, more maths. And, all at once, a vision returned to her of her 7-year-old mind's eye: numbers existing in space, in different shapes and colours, easily read and connected and related. To find the answer, all she had to do was consult her mental picture and read it off.

And when she moved up to a higher class, this private world of numbers was lost. Rudely pulled outside her comfort zone, she lost track of it. There was pressure to adopt newer, more abstract methods immediately and her confidence evaporated.

This friend is a professional painter. For her, seeing is everything. She is an extreme case, probably. But seeing does count for a lot, and being able to see numbers, and number problems, is a key skill of the successful mathematician.

In the early years, children build their concept of numbers in a very concrete way. They use numbered blocks, counters, toy digits and so on. The senses of sight and touch are heavily involved. Gradually, we wean children off these aids and get them to work with digits on paper. That's fair enough, because it is a cleaner, faster way to the answer, and common to everyone. But the visual/spatial model in our minds is more important. It can be the foundation for our abstract thinking. For example, Japanese children trained to use an abacus at super speed will close their eyes and picture an abacus, or mime the movements of counting on one, to help with super-fast mental arithmetic.

To learn about the world we have to organise it in our minds, and that includes numbers. One way to build up the mental world we need is to see numbers in sets.

The first set of connections we all learn is 1, 2, 3, 4 … After that we can learn 2, 4, 6, 8 … and 10, 20, 30, 40 … and, crucially, we eventually notice connections. For example, I see that 25 is in lots of sets: {5, 10, 15, 20, 25, 30 …} and {25, 50, 75, 100} and {4, 9, 16, 25, 36}. When we think of 25, we should call to mind all the sets to which it belongs. Of course, these are actually limitless, and lines and links and pairs and groups sprawl in every direction from every starting point.

Our brains explore numbers in this way just for the fun of it. But it is also a really important part of number competency.

By seeing each number as being in various different sets we start to give it a character, almost like a person or place, and this helps us to build our picture of the number system – a picture that we need to consult like a map that we have memorised. If the numbers from one to one hundred are a village, then the more different routes we learn through it, associations we make and patterns we can spot, the more at home we are. Without the mental map, we are like people in a city who only feel comfortable sticking to the big main roads, and ignore the short cuts and connecting side streets that make life easier.

For people who don't see connections, 12 is just 12, sitting still between 11 and 13. They don't want 12 to be 10 + 1 + 1 and 5 + 5 + 2 and 4 x 3 and 3 x 4. They don't want 10 + 1 + 1 to be 1 + 10 + 1 and 1 + 1 + 10. Thinking this way leaves them all at sea, as if the numbers are just disintegrating. And, of course, we all feel like that sometimes if people show us things too fast. So, this kind of work should be done with things we already know quite well.

Do what you can at all times to show that each number has infinite connections with every other number, that it can be made of numbers bigger and smaller than itself in any number of ways, and stand in all kinds of sets and sequences. And do it by showing that, crucially, this is half the fun of numbers.

All these connections are far easier to see than to comprehend through the written word. What we are going to do here is to bring those sets and sequences to life.

There are plenty of different ways to do it, but getting kids to *be* numbers is great fun and can help them to internalise the relations between numbers, strengthening their mental grip on the whole thing. As ever, the children who are struggling will have no strategy – no way to 'see' numbers.

# Things to Do 25

**Do** Give every child a number.

You can just count round the group and give them a number each and start. But, better still, give them a piece of paper on which they write the number they have been given, even decorating it a bit – this is all about *personalising* or *personifying* the numbers. They can hold or wear their number and keep that number for anything from a few minutes to a few weeks.

**Do** Get them into a circle.

**Say** Stand up when your number is included.

Ready? Stand up if you are an even number.

Next, all the children whose number is in the 2 times table must stand up, and the others stay down. Use the instructions below and add some of your own. If you want you can get them doing some mental arithmetic: 'Stand up if you are four away from ten!' or '… next to a number in the 3 times table!' And, of course, sometimes no one should stand ('… half an odd number') and sometimes everyone ('… half an even number').

**Say** Point at someone if you think they should be standing … or should be sitting.

As you give more difficult instructions, some children will get it wrong. Ask those who are pointing to explain why they think their classmate should change, and see if they influence anyone.

Here are my ideas, and I'm sure you can come up with some cute ones of your own. These are in three groups, in ascending order of difficulty:

Stand up if …

- ❖ You are the highest number in the group.
- ❖ You are in the 2 times table.
- ❖ You are an odd number.
- ❖ You are a one-digit number.
- ❖ You have the digit 1 in you.
- ❖ All your digits are made of only straight lines.
- ❖ You are on a clock face.

❖ You are the age of a child at this school.

❖ You are the middle number in the group.

❖ You are a prime number.

❖ You are a square number.

❖ You are in the 13 times table.

❖ You can be divided by four without a remainder.

❖ You are half of an even number.

❖ You are one less than a multiple of ten.

❖ You are double an odd number.

❖ You are the number of players in a sports team.

❖ Multiplying you by three makes an odd number.

❖ You are a factor of sixty.

❖ Your percentage – for example, if you are 5 then 5% – can be simplified to a lower fraction.

❖ You are the sum of two primes.

❖ Adding your digits together makes more than eight.

❖ You are the number of legs on an animal.

On a snow-struck day, or when a trip takes out half of the school, bung two classes together in the hall, play this, and the other teacher can do some planning over a cup of coffee. In a bigger space, like the hall, there are more possibilities. You could give half the children operation symbols and equals signs while the others have numbers. Set the challenge that the children have to form groups to make complete sums. For example, the children who are 2, +, 5, = and 7 can form a line. Those children in partly formed groups who haven't made a successful sum have to split up and re-mingle until everyone finds a place in a sum somewhere. You can make it competitive by saying that the ones left at the end are out of the game, then everyone has to find new partners. It will be noisy, obviously.

Rather than give out operation symbols, you could call out an operation and everyone has to form groups of at least three where two of them, with that operation, form the third. So, if you call out 'plus' then 4 and 3 could team up with 7 (as they could if you called out 'minus'), and if you called out 'times' then 4 and 3 could team with 12.

Alternatively, don't bother with the operation symbols and rely more on imagination: get the children to form groups of 'friends', so if 12 and 4 get together they could form a group of friends with 3 (because 3 is 12 ÷ 4), 8 (which is 12 − 4) or 16 (12 + 4). Again, those with no 'friends' are out. Bear in mind that those children most inclined to hook up with their real friends are likely to be out fairly soon, because they won't be using the right criteria, so it should cut across their friendship groups. Of course, if you don't like the idea of the friends/no friends labels you could easily rename them, but I think there is something helpful in seeing numbers that fit together as friends.

Fractions too can be illustrated with human figures in a number of ways. One is for you to call out a fraction: 'Three-sevenths!' The children need to form a line with three standing and four kneeling. Do it without selecting anyone to come forward, so that they have to organise themselves, stepping out if there are already enough people. The ones who did that fraction can step aside for the next one, until all the children have had a turn, and you start again.

# Key Word 12: Kinaesthetic

This word combines the idea of kinetic, which is to do with movement, and aesthetic, which is to do with the senses. As I said in 'Squintasticadillion and One', you don't have to subscribe completely to any theory about learning styles to see that adding movement and sensory experience to a lesson will usually improve its impact. And there is plenty of evidence that a brain activated in these ways is more ready to learn.

In the activity above, when you have a whole line of children standing and sitting to illustrate the 3 times table, or the square numbers, patterns are easy to spot and remember. And because there is the challenge of listening and moving when required, it allows you to go back over old mathematical ground without boring the children who are already comfortable with it.

Primary school children are pretty willing to move around and to be looked at by their classmates. With adolescents this can be a problem. The simple fact that your body is increasingly mature and unfamiliar to you can make you reluctant to draw attention to it by standing up, especially if it is already the subject of other people's remarks. So, go easy on self-conscious people and allow them to join in when they are ready.

There are lots of simple ways to make a lesson a little more kinaesthetic. Here are a few:

- ❖ Children come to the board and write on it.

- ❖ Write words or ideas on pieces of paper that the children can pick up and move.

- ❖ The teacher stands in a different place in the room when talking to the class or sits on a different chair.

- ❖ Make lines out of thin strips of paper, or use rulers, so that the children can form shapes to discuss.

# Stairs Into Space

Apparently, some people on the autistic spectrum like to count. The world around them is swollen with opportunities for calculation. On the shelf of the corner shop they see 8 packs of playing cards for £1.49 each. So they might start by noting to themselves that (because 8 ones plus 8 halves makes 8 + 4, which is £12 minus 8 x 1p) all 12 would cost £11.92. They might go on then to calculate that with each pack containing 4 kings, there are 32 kings on the shelf, but that wouldn't detain them for more than a fraction of a second before they wondered how many hearts there are on the shelf. Not including picture cards, you would have $10 + 9 + 8 + 7 + 6 + 5 + 4 + 3 + 2 + 1$ (the ace) = 55 hearts in each pack, so 440 hearts in 8 packs. But most packs have 2 small hearts in diagonally opposite corners of each card next to the digit. So, if you're looking at a pack like that it gives you another 20 hearts for those 10 cards mentioned before, and then you would need to include the picture cards, which actually have 4 hearts (a big and a small one in both opposite corners), so that is another 12, bringing the total to 472.

In this book I spend quite a lot of time talking about purpose. But what is striking about that last train of thought is that it is so utterly divorced from any purpose at all – you don't need to know how many hearts there are, and why would you care? One of the challenges for people who do like to calculate for the sake of it is to accept that most other people don't.

But there are things 'normal' people commonly do that lack purpose in the same way. A Sudoku, for example – it's a waste of time and brain power. A crossword. A quiz full of trivia. None of these have any purpose but our own amusement. Looked at this way, it does seem kind of tough on people who have been diagnosed as on the spectrum to be told that their happy journeys into pointless facts and figures are weird, whereas the pub quiz and the newspaper crossword are normal. The standard explanation is that autistic/Asperger's behaviour is extreme, and that is the problem. No doubt that is true, but their thinking can also be very original. Instead of solving problems set by others, they are devising and solving their own.

I am definitely not saying that children would be better off if they were a bit more autistic. But I would say that, despite all the problems that autistic thinking creates, we might learn something from it too. And that is to try to lift the lid off the physical world and peer at all the numbers wriggling underneath.

Most children, to some extent, do retain a zeal for pointless calculation, especially where there is a bit of competition or challenge involved. You can use this to encourage them to look at the world around them and make guesses and calculations. They can have a lot of fun and it helps them to notice how numbers are embedded in everything. So, start looking around the school environment – and beyond – for little games and challenges, and get them to do the same. Here are some examples:

❖ How many windows are there in the school?

❖ How many computers are there in the school?

❖ How many children are there in the school? How many boys? How many girls?

❖ How many keys are there on the computer keyboard? Or the piano keyboard?

❖ How many squares are there in the fence around the car park?

❖ How many seconds are there in a year?

❖ What percentage of our lives are we awake?

- ❖ What percentage of our waking hours are we at school?

- ❖ Are there more countries in Africa or South America?

Here is where we part company, perhaps, with a more autistic way of thinking, because we are going to pay a lot more attention to estimation. Along with the practical maths involved in finding the answers, children can use these little questions to improve their assessment and mental arithmetic, and also to verbalise their thinking in order to persuade others. Rather than just processing pure numbers, the children are honing their common sense and communication.

# Things to Do 26

**Do**  Choose a flight of stairs and count them.

Measure vertically from the ground to the top step.

Most schools have a flight of stairs somewhere. Preferably choose a set that the children use every day, and preferably with more than five stairs. It may also help if it is not the main thoroughfare of the school, in case you want to take the class out and have a look at some point. Incidentally, it doesn't have to be the whole stairway – you can use just the flight between the ground and the first landing.

Before the lesson, measure and note the distance from the ground to the top of the last step (i.e. the height covered by the stairs). Then count the stairs. Your whole lesson is based on these numbers.

Take out the tape measure and tell the class that you have measured the height from the ground to the top of the last stair and write up the measurement. Then you have one simple question:

**Say** How many stairs are there?

I guarantee that no matter how many times they have gone up those stairs they will not have counted them, although a few might claim they have. For this task, the children need to use a bit of mathematics and some common sense to get an answer. One way to get away from sheer guessing and on to reasoning is to say:

**Say** What do you think are the lowest and highest possible numbers of stairs?

If needed, throw in some numbers yourself. Could there be just one stair? What about forty? Narrow it down.

Now, let's say the total height you gave them was 280 cm. The class may not express it in these terms, but they are now looking at: stair height x number of stairs = 280. So, to know the number of stairs they need to know the height of each one. Again, encourage the children to come up with a realistic range using a ruler. They may conclude that the stairs are definitely higher than 5 cm each but definitely can't be more than 25 cm. Perhaps they can agree to narrow it down further.

The children may debate whether the stairs are all the same height. If they do, this is a good chance to bring in some internet research (I do mine in advance, but it might be part of your plan to get the class to do it). Basically, there are various regulations that say stairs in the same flight must be the same height (no stair can be more than 1 cm higher than any other).

Now comes another opportunity to feed in more data. Internet research throws up contradictory results but there is some consensus: the maximum stair height is 22 cm and the minimum 15 cm. Different heights are used in different types of buildings. Can the children think where higher steps are allowed and lower steps are better? What would be best in a school for children of their age?

The aim should be to reason their way to a likely answer and then test it out. Agreement from everyone isn't necessary before they test it out. If there are twelve stairs, and a child estimated eleven or thirteen, they should be pleased with that result. They simply don't have enough information to predict the exact number. The idea is for them to accept that they can't hit the bullseye (if they do, it's just luck) but use an increasing amount of information to hone their estimate to a more accurate range.

At the end you could send a couple of kids out to go and count, but it might be better to let them check after the lesson. Hopefully, you will have generated enough interest that they will head off to do it anyway. Having said that, I finished one class in this way and then went down to the office to sign out. Two girls followed me, sent by the teacher on some other errand, down the very stairs we had just spent half an hour talking about. 'So how many were there?' I asked. 'How many what?' they said, looking puzzled. 'Stairs!' I reminded them. 'Ooooh!' They went back to check. 'Is it ten?' I heard them call, as I walked off towards my bike.

# Things to Say 12

*'Test it!'*

With any line of enquiry, there is always the danger that we will discover that it leads nowhere. So, when facilitating an enquiry it is tempting to close down certain discussions before they begin because you know already they are barking up the wrong tree.

But resist! We all need to learn that a successful project rarely consists only of successful steps. Disappointing though they are, these setbacks are part of the process. Children find it particularly difficult to see this. If children are given a puzzle that requires a certain amount of trial and error to solve, try asking them afterwards what helped them to do it. They will remove all the 'error' part of the process and only tell you about the test that finally came up with the result. They fail to see that you can't do only successful experiments – they wouldn't be experiments if you did.

By allowing children to try out ideas and go on to dismiss them, you are getting them accustomed to the idea that these 'failures' are not failures at all, but successful attempts to discover whether something works or not – in this case, not. They are disproving possibilities and working towards the answer by elimination – and sometimes that is the only strategy we can have.

Of course, at the other extreme, letting groups wander way off in the wrong direction completely with no guidance will not be good for them. So, like everything else, it is a question of finding the balance, and you can't always get it right.

But don't be deterred by the most common objection to letting a group pursue an avenue to a dead end: that they are losing valuable time – an increasingly precious commodity in the modern classroom.

If, by following leads to dead ends and then revaluating, children learn to persevere, adapt methods and work systematically with few clues, they are in possession of a valuable skill – one that will help them in the classroom, the workplace, the laboratory and the examination hall. Is that a waste of time?

So, when you hear an answer, don't automatically declare it right or wrong. Where you can answer, 'Test it!'

# Bone Setting

The word algebra comes from Arabic – *al-jabr*. The al- part actually means 'the' in Arabic and has been carried into quite a lot of phrases we have inherited from that language: alcohol, albatross, Alcatraz, alchemy, alcove, algorithm.

*Al-jabr* were the first two words in the title of a treatise on the solving of equations, translated into Latin in the Middle Ages. Literally, *jabr* meant completing, or restoring, or the setting of a bone after it has been broken. It reminds me of the brokenness of a jigsaw puzzle which we complete, or reset, and which gives much the same pleasure as solving equations.

In this chapter, I will show you a way of exploring algebra with kids just for the fun of it. But I would also like to say a bit about what algebra is, and describe how it can be used in a practical setting. I think this could be useful because it might have been missing from a lot of teachers' experience of algebra at school.

So what is algebra? The basic answer is: it is a method for representing how one number relates to another, even when they both change. For example, whenever my age changes so does my sister's: she will always be two years younger than me. Algebra gives us a way to write this relationship between the two numbers, whatever they may be at any particular time:

Andy's age = sister's age + 2

(It's the same the other way round: Andy's age – 2 = sister's age)

Instead of the whole words we could use A to mean Andy's age, and S to mean sister's age, giving us:

$A = S + 2$

Mathematicians often use the letters X and Y, so they might prefer Y = X + 2, but you can use any symbols you want. Just remember that the reason we are using symbols is because the numbers behind them could be anything – all we know is that there is a relation between one and the other. The magic is that if you know one of the numbers, you can work out the other. So, if you know my age then you can work out my sister's, and vice versa. This is exactly what I would do if someone asked me how old my sister was. I wouldn't know off the top of my head, so I would take my age (which I usually can remember) and take away two.

The key thing to remember when working with equations is that they are called equations for a reason – because right in the middle is an equals sign. This sign says that whatever is on one side is equal to what is on the other. If the number on the left goes up, so does the number on the right – just like the ages of me and my sister. This gives us something useful to play with when we are looking for answers because we can manipulate the equation. To find my sister's age from A = S + 2 I do two things: *reverse the calculation (so subtract to get rid of the plus sign)* and do that to *both sides.*

Working with slightly more complex equations, nothing changes. So, let's say the number you want to know is surrounded by a few calculations, like this:

$T = M + £10 \times 2$

In this equation, M stands for 'my debt' and T for 'toaster'. And this is the situation it represents: my flatmate is buying a new toaster for the flat. I would owe him half the cost of the toaster but he owes me a tenner. So that has to be deducted before I pay him. At the moment, you can't see how much M is worth because it has got all these calculations around it. So, what you want to do is to move all the calculation bits away so that M is standing on its own. And you do this how? By *reversing* those calculations and doing it to *both sides*. So, divide by 2 (which reverses x 2) and take away 10 (which reverses + 10). So you get:

$$T \div 2 - 10 = M$$

This says that to find out what I must pay, I take the cost of the toaster, divide it by 2, and take away 10. So, as soon as I know what the toaster costs, I can do the maths.

At this point, you would be entitled to object that you can do all this a lot more easily without algebra. And that is true. As long as you can do all the sums in your head, algebra is of no use to you. But the important thing with these basic sums is to appreciate that, regardless of its usefulness so far, algebra does work for them. Where it comes into its own is when you can't do those sums in your head. Because then you can trust the principles you have just seen to solve a whole blackboard of scrawl. Or you could … in theory.

Algebra is also what we need to express general rules, regardless of particular numbers. So, going back to the age example again, $S = A - 2$ means 'However old Andy is, his sister is two years younger', and expresses something true no matter what our ages happen to be at any one time. Likewise, when we use equations in physics they allow us to describe general rules without needing to talk about actual amounts. So, going back to Einstein's $E = mc^2$, this equation defines the relation between energy and mass (it is the amount of mass multiplied by the speed of light squared) in general – something that can only be done if you allow letters, or other symbols, to stand in for whatever the numbers turn out to be in any one situation.

Numbers can only tell you specific facts about the world. Algebra allows you to reveal the laws that govern the world. However, all we are going to use it for in the next section is to play.

# Things to Do 27

**Do**  On small pieces of paper, write simple mathematical instructions (like + 2, x 3, –3 and so on) on each one.

Choose the simplest one to start with – let's say it is + 2. Place the piece of paper face down in front of you. (Here is where you have to make sure what you have written isn't visible through the paper!)

**Say**  I want you to work out what is on this piece of paper. Here's how you do it. You give me any number you like and say, 'X is …' and the number – for example, you can say, 'X is 4'. I will then give you another number.

Someone give me a number. X is … ?

At this point, the class will probably not be 100% sure what you mean, but that doesn't matter because someone will still give you a number. You then add 2, in accordance with what it says on the paper.

**Say**  Y is …

Keep asking them for numbers, so they say, 'X is …' and you reply, 'Y is …', and before long they will start to work out that you are adding two each time. When this rumour has worked its way round the class a bit, you can turn over the paper to reveal that they were exactly right.

From now on, get someone to record the values of X and Y in two columns, so that the children can see which Xs have already been said and what Y was in each case. You could end up with something like this:

| X | Y |
|---|---|
| 4 | 12 |
| 8 | 24 |
| 1 | 3 |
| 100 | 300 |

As you can see this table is for x 3, or to give it its full title: Y = X x 3.

**Say** Now you are going to play the same game in pairs. Secretly write down what you are going to do – and remember you have to do what it says on the paper each time, whatever number your partner gives you. If it is too hard, just say, 'Too hard!' and that is a kind of clue.

**Do** Get one pair to play each other in front of everyone else to check/demonstrate, then they can all work in pairs.

**Say** Now everyone try. When your partner works out what you are doing, swap round.

Leave them to play for a while, but go round the pairs and check to see who is struggling with the rules. See if anyone is recording the results – draw attention to it if they are because that is good thinking.

**Say** Stop! Now, has anyone got quite a difficult one that their partner couldn't get for us all to try?

This activity will settle at the children's natural level. They will make it harder and challenge each other. The next stage to introduce (and children rarely think of it themselves) is to complicate the equation by adding a second operation. Take a look at this table.

| X | Y |
|-----|-----|
| 7 | 12 |
| 15 | 28 |
| 10 | 18 |
| 2 | 2 |
| 1 | 0 |
| 100 | 198 |

This will fox quite a few people. And that is because it has two operations: $Y = X \times 2 - 2$. Notice that certain values of X, such as 1 and 100 can be quite revealing. Elicit this from the students:

**Say** Which X numbers are very useful for helping us to work it out? Why?

# Things to Say 13

*'What does your partner think?'*

Language teachers have put a lot of thought into how to get people talking. In their case, the barrier for their students is that they are trying to talk in a foreign language. The barrier in maths is … well, yes, for some it is like a foreign language. But whatever the subject, most people are nervous or shy about talking in front of the whole class. And interacting directly with the teacher in front of one's peers is intimidating for most.

In language teaching, pair work is used extensively to develop speaking skills. In maths, with pair work, students can:

❖ Build confidence before speaking to everyone else.

❖ Iron out misunderstandings.

❖ Rehearse and develop ideas.

❖ Hear an opposing point of view to challenge theirs.

❖ Hear a similar point of view to reinforce theirs.

If you give children the chance to discuss in pairs, you will get a response when you go back to the whole group and hear feedback. During the talking time you can also be going from pair to pair, picking up ideas of what is being discussed, understood or misunderstood. You can also boost the confidence of individuals and encourage them to share their idea with the whole group when the pair work is finished. In my opinion, pair work is best used as a stage in building up to a whole-class conversation; pair work without feedback is often a waste of time.

In the algebra activity, the pairs are given specific instructions for a game. I have recommended that before the whole class divides into pairs, you choose one pair to play in front of everyone else. If they haven't understood your wonderful instructions, you then get the chance to fix that before the whole class descends into confusion!

# Voyager[1]

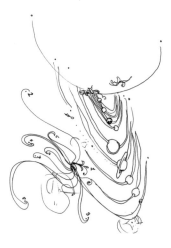

How would we speak to aliens? If we ever came face to face, then presumably there would be some kind of gesture or movement possible, from which we could build up a mutual understanding, just as if we came across someone in the jungle who had never laid eyes on an outsider.

But what if the aliens were on their planet and we were on ours?

Carl Friedrich Gauss, one of the greatest mathematicians of the nineteenth century, suggested that a giant right-angled triangle be created on the Siberian tundra, using ten-mile wide strips of pine forest. This would be visible to any creature looking at Earth through a telescope and would convince them that there was intelligent life here. Others,

---

1  This session was inspired by J. B. Nation's paper, 'How Aliens Do Math'.

supposedly, proposed digging an enormous canal in the Sahara, filling it with water and burning kerosene on the surface to send light signals in various shapes.

On 5 September 1977, the *Voyager 1* space probe was launched from Cape Canaveral. Its primary purpose was to get close to the other planets in the solar system and send back data and images – which it has done spectacularly. While they were at it, NASA included some messages to alien life forms, knowing that *Voyager 1* would one day plunge into interstellar space and begin a long journey through the emptiness towards other stars – perhaps other life forms.

What did they put into the capsule? Two golden records that would allow their discoverer to play back various pieces of information, music (from Mozart to Chuck Berry), assorted diagrams of the solar system, depictions of the structure of atoms, and figures of a male and female human being. There were some worries that permission might not be given to include the messages if the figures were nude, so they were in silhouette. It is hard to imagine that any aliens would have opened the capsule, seen the naked man and woman, and muttered, 'Ugh, what a dirty lot …'

Throughout the 1970s there was a lot of interest in exactly what form our communications with extra-terrestrials should take. The aliens may well not rely on their sense of sight in the way we do, so pictures might not work. The same goes for sound. Various electromagnetic signals, such as radio wave pulses, could perhaps be picked up by aliens' equipment. Unless, of course, they perceived these parts of the electromagnetic scale directly – as we do with light, which would be even better.

So, what is needed is something that can be broadcast, or recorded, in any number of forms to make it as likely as possible that aliens will notice it and attempt to decipher it.

In the film *Close Encounters of the Third Kind*, scientists communicate with aliens that land on Earth by playing a melody, which the aliens play back.

But mathematics may be the key. Carl Sagan worked on the messages for *Voyager 1* and wrote a science fiction novel, *Contact*,[2] later turned into a film, which imagines contact with alien life forms. He felt that any intelligent being would have mathematics and so would recognise representations of the prime numbers.

Is he right that intelligent aliens would have to understand maths? As we have seen in this book, counting can be done with different words in different languages, and with different symbols, but most of us agree that when I say 'five' and my Japanese friend says '*goh*' we are both talking about the same number. So, while it is possible to imagine a life form, or perhaps an entire civilisation, that had no mathematics, it is impossible to imagine a life form that had a *different* maths.

And that could make maths unlike every other field of knowledge, because it is possible to imagine a different physics, with no objects or friction or gravity. Physicists even tell us that to describe the universe we need to understand that it has many more than three dimensions – perhaps ten, or eleven or twelve.

If there are other thinkers out there, the one thing they may share with us is an appreciation of numbers, so maybe they would recognise prime numbers. Maybe they too have to go to school and learn times tables. One of them may even be writing a book about how to make those lessons more interesting and fun.

---

2 Carl Sagan, *Contact: A Novel* (New York: Simon & Schuster, 1985).

# Things to Do 28

**Do**  Tell some of the stories above or describe Gauss's or Sagan's ideas.

Play a clip from one of the films.

Or just launch straight into the questions:

**Say**  What message could we send to aliens on another planet to show that there is intelligent life here?

Would they recognise mathematical messages? Which ones?

How could we make them easier to recognise?

# Bibliography

Bellos, A. (2011). *Alex's Adventures in Numberland* (London: Bloomsbury, 2011).

Boaler, J. (2010). *The Elephant in the Classroom: Helping Children Learn and Love Maths* (London: Souvenir Press).

Du Sautoy, M. (2010). *The Number Mysteries* (London: HarperCollins).

Eastaway, R. and Wyndham, J. (1998). *Why Do Busses Come in Threes?* (London: Robson).

Enzensberger, H. M. (1998). *The Number Devil* (London: Granta).

Gullberg, J. (1997). *Mathematics from the Birth of Numbers* (London: W. W. Norton).

Heath, C. and Heath, D. (2011). *Switch: How to Change Things When Change is Hard* (London: Random House).

Kasner, E. and Newman, J. R. (2013 [1940]). *Mathematics and the Imagination* (New York: Dover).

Markowsky, G. (1992). 'Misconceptions about the Golden Ratio', *College Mathematics Journal* 23(1): 2–19. Available at: <http://www.math.nus.edu.sg/aslaksen/teaching/maa/markowsky.pdf>.

Nation, J. B. (2003). 'How Aliens Do Math' (University of Hawaii). Available at: <http://www.math.hawaii.edu/~jb/four.pdf>.

Ollerton, M. (2003). *Getting the Buggers to Add Up* (London: Continuum).

Sagan, C. (1985). *Contact: A Novel* (New York: Simon & Schuster).

Stewart, I. (1995). *Nature's Numbers* (London: Weidenfeld & Nicolson/Orion).

Stewart, I. (2012). *17 Equations That Changed the World* (London: Profile Books).

Tammet, D. (2012). *Thinking in Numbers: How Maths Illuminates Our Lives* (London: Hodder & Stoughton).

Wittgenstein, L. (2009 [1953]). *Philosophical Investigations*, 4th edn (Chichester: Wiley-Blackwell).

Worley, P. (2010). *The If Machine: Philosophical Enquiry in the Classroom* (London: Continuum).

Worley, P. (2011). 'Il Duomo', in *Once Upon An If* (London: Bloomsbury), pp. 151–155.

Worley, P. (ed.) (2012). *The Philosophy Shop* (Carmarthen: Independent Thinking Press).

Worley, P. and Day, A. (2012). *Thoughtings: Puzzles, Problems and Paradoxes in Poetry to Think With* (Carmarthen: Independent Thinking Press).

# Acknowledgements

The team at Crown House for backing this project.

Tamar Levi for illustrations that capture the spirit of the text as intended.

Peter and Emma Worley for setting up The Philosophy Foundation and introducing me to work in the classroom. Also my varied colleagues at The Philosophy Foundation for being stimulating and enjoyable company and providing all kinds of ideas, many of which I may have unwittingly passed off as my own in subsequent conversations.

Alison Sharp and Kirstin Fisher at Ravenscroft School for pushing me so far out of my comfort zone that it disappeared behind me in the distance. Also, the teachers there for allowing me into their classrooms and giving me the benefit of their own experience.

All the children and teachers I have worked with over the last few years who have shown me where these ideas were going right and wrong.

Dilip Sequeira for encouraging feedback and excellent notes.

My wife, Shalini Sequeira, for unstinting support and *highly* intelligent advice.